EXPOSED

This book is for Roy, who was the first. For Wendy, who wept on my shoulder as she remembered the babies she'd lost decades before. For Shirley who rang me every week, asking: 'What are we doing next, dear?' And for Barry, who told me as he lay dying: 'Don't let them get away with it.'

It is for all those who were unheard, and who died before they could stand at ease. The men who cried as they told me of things they could not tell their wives, the widows who raged at official lies, the children with questions that went unanswered.

They gave me a lever, and a place to stand.

EXPOSED

The SECRET HISTORY of Britain's Nuclear Experiments

SUSIE BONIFACE

Pen & Sword
MILITARY
AN IMPRINT OF PEN & SWORD BOOKS LTD.
YORKSHIRE – PHILADELPHIA

First published in Great Britain in 2024 by
Pen & Sword Military
An imprint of
Pen & Sword Books Ltd
Yorkshire – Philadelphia

ISBN 978 1 03612 100 6

A CIP catalogue record for this book is available from the British Library.

Typeset by SJmagic DESIGN SERVICES, India.

Printed and bound in the UK by CPI Group (UK) Ltd.

Pen & Sword Books Limited incorporates the imprints of Archaeology, Atlas, Aviation, Battleground, Digital, Discovery, Family History, Fiction, History, Local, Local History, Maritime, Military, Military Classics, Politics, Select, Transport, True Crime, After the Battle, Air World, Claymore Press, Frontline Publishing, Leo Cooper, Remember When, Seaforth Publishing, The Praetorian Press, Wharncliffe Books, Wharncliffe Local History, Wharncliffe Transport, Wharncliffe True Crime and White Owl.

For a complete list of Pen & Sword titles please contact

PEN & SWORD BOOKS LIMITED
George House, Units 12 & 13, Beevor Street, Off Pontefract Road,
Barnsley, South Yorkshire, S71 1HN, England
E-mail: enquiries@pen-and-sword.co.uk
Website: www.pen-and-sword.co.uk

or

PEN AND SWORD BOOKS
1950 Lawrence Rd, Havertown, PA 19083, USA
E-mail: uspen-and-sword@casematepublishers.com
Website: www.penandswordbooks.com

FSC
www.fsc.org
MIX
Paper | Supporting responsible forestry
FSC® C013604

Contents

ACKNOWLEDGEMENTS

JOURNALISTS WORLDWIDE HAVE spent decades alerting others to what befell thousands of people affected by the British nuclear weapons programme, and I am indebted to them. Thank you in particular to my predecessors at the *Mirror*, Paul Foot, Richard Stott, and Alan Rimmer, and to my editors Tina Weaver, Alison Phillips, and Caroline Waterston, who gave me the time and space to investigate.

For further reading I recommend *Grappling With The Bomb* by Nic Maclellan, about Pacific Islanders, and *Maralinga* by Frank Walker on Australian tests and veterans. Joan Smith's *Clouds of Deceit* covers the early years of the UK campaign. Thanks to all three for letting me quote their work, and to the family of Flight Lieutenant Joe Pasquini for his recollections from *Flying Between Two Suns*. I tip my hat to Dr Elizabeth Tynan, who wrote *Atomic Thunder: British Nuclear Testing in Australia*.

They would all join me in paying tribute to the nuclear families who have fought for the truth for a lifetime. Doug Hern's collection of papers, which he kindly shared with me over the years, and Mike Chainey's continuing efforts at the National Archives, have been invaluable.

There isn't space to list and honour all those I have interviewed over almost two decades. The nuclear community has always been honest, dignified, and good-humoured, while enduring and telling me of things that have scarred them deeply, and I thank them for that. I have tried to be kind to the scientists involved in the design of these ferocious weapons, and to explain why they were, and are, felt necessary. William Penney, the mathematician who was 'Father of the Bomb', was not responsible for the scandal that came in its wake; he is our Oppenheimer, and in some ways a victim too.

People always ask 'who's to blame?' There is only one answer: whoever is currently Prime Minister. They alone have the power to end the longest-running scandal in British history. I have done my best to shout about it to the last six people who held that office, with varying success. Maybe the next will do the right thing.

In parts I have written about my own involvement with the campaign, an awkward thing to do, and have done so in the third person: not to confuse the reader, but to report on my own actions with the same attitude as I do others.

There are three people without whom this book would not exist: my wonderful dad, who taught me about radiation and the principle of leverage, my fabulous mum who has listened to years of grumbling about people she's never met, and my beautiful daughter, who has so often had to share me with the veterans, and not complained (much) about doing so.

Thank you for teaching me that whatever is broken can always be fixed.

TIMELINE

1945 US fires first atomic weapon

1946 First resolution of United Nations calls for end of atomic weapons

1949 USSR fires first atomic weapon

1952 UK fires first atomic weapon; US fires first hydrogen bomb

1955 USSR fires first H-bomb

1958 Campaign for Nuclear Disarmament holds first meeting; UK fires first H-bomb

1959 Nuclear tests banned in Antarctica

1960 France fires first atomic weapon

1961 USSR fires 50-megaton Tsar Bomba

1963 US, UK and USSR limit testing in space, underwater or in the atmosphere

1964 China fires first atomic weapon

1967 Latin American nations sign nuclear-free treaty

1968 UK, US and USSR sign non-proliferation pact

1974 India conducts first atomic test

1979 Nuclear test in Indian Ocean conducted by South Africa with Israeli help

1982 1million people rally in New York against nuclear weapons

1985 South Pacific becomes nuclear-free zone

1986 Whistleblower reveals Israel has up to 200 nuclear warheads

1991 Last British test, Operation Bristol, conducted

1995 South East Asia becomes nuclear-free zone

1996 Africa becomes nuclear-free zone; Ukraine follows; China, France, UK, US and Russia sign comprehensive nuclear test ban treaty

1998 India and Pakistan conduct underground atomic weapons tests

2006 North Korea fires first atomic weapon

2015 North Korea claims first hydrogen bomb

2020 UN Treaty on the Prohibition of Nuclear Weapons ratified, making them illegal in international law. US, Russia, China, Britain, and France refuse to ratify it

UK Nuclear Weapons Tests, 1952–1958

1 kiloton (kt) = 1,000 tons of TNT equivalent
1 megaton (mt) = 1,000,000 tons of TNT

Codename	Location	Date	Yield (kt)	Type
HURRICANE	Monte Bello	October 3, 1952	25	ATOMIC
TOTEM I	Emu Field	October 14, 1953	10	ATOMIC
TOTEM II	Emu Field	October 26, 1953	8	ATOMIC
MOSAIC I	Monte Bello	May 16, 1956	15	ATOMIC
MOSAIC II	Monte Bello	June 19, 1956	60-98	BOOSTED FISSION
BUFFALO One Tree	Maralinga	September 27, 1956	15	WARHEAD
BUFFALO Marcoo	Maralinga	October 4, 1956	1.5	WARHEAD
BUFFALO Kite	Maralinga	October 11, 1956	3	WARHEAD
BUFFALO Breakaway	Maralinga	October 21, 1956	10	WARHEAD
GRAPPLE 1 Short Granite	Christmas Island	May 15, 1957	300	H-BOMB FAILURE
GRAPPLE 2 Orange Herald	Christmas Island	May 31, 1957	720	H-BOMB FAILURE
GRAPPLE 3 Purple Granite	Christmas Island	June 19, 1957	200	H-BOMB FAILURE
ANTLER Tadje	Maralinga	September 14, 1957	1	ATOMIC
ANTLER Biak	Maralinga	September 25, 1957	6	ATOMIC
ANTLER Breakaway	Maralinga	October 9, 1957	25	ATOMIC
GRAPPLE X	Christmas Island	November 8, 1957	1800	HYDROGEN
GRAPPLE Y	Christmas Island	April 28, 1958	3000	HYDROGEN
GRAPPLE Z Pennant	Christmas Island	August 22, 1958	24	ATOMIC
GRAPPLE Z Flagpole	Christmas Island	September 2, 1958	1000	HYDROGEN

Codename	Location	Date	Yield (kt)	Type
GRAPPLE Z Halliard	Christmas Island	September 11, 1958	800	ATOMIC
GRAPPLE Z Burgee	Christmas Island	September 23, 1958	25	ATOMIC

Note on units:
Roentgen (R) is an historic measurement unit of gamma radiation
REM, or REP, estimates how much R is absorbed by humans
Sieverts or miliSieverts (mSv) is a modern estimate of genetic damage
Wherever possible this book relates doses in mSv. The average annual background dose of radiation in the UK at the time of the tests, from natural and man-made sources, was around 2 mSv. This equates to roughly 0.2 R, or 0.2 REM.

The Minor Trials, 1953–1963

Codename	Location	Date	Purpose	Number of blasts
KITTENS	Emu Field	Sept/Oct 1953	Triggers	5
TIMS	Maralinga	1955–1963	Cores	321
KITTENS	Maralinga	1956–1961	Triggers	99
RATS	Maralinga	1956–1960	Shockwaves + compression	125
VIXEN A	Maralinga	1959–1961	Fires	31
VIXEN B	Maralinga	1960–1963	Accidents	12

UK troops also served at Operation DOMINIC, a series of US weapons tests held at Christmas Island between April and July 1962. A total of 31 weapons were exploded in 78 days, with a combined yield of 38,000 kilotons. The stated purpose was testing design and reliability.

Further troops took part in clean-ups. Maralinga saw AYRES in 1960, another in 1963, HERCULES in 1964, and BRUMBY in 1967. Plutonium was removed from the site in 1979, with a survey in 1981 (MCEWAN) and another in 1998 (WOODVILLE). An Australian government clean-up began in 2000. At Christmas Island, there was a AWRE survey in 1964 (OLDBURY), another in 1969, clean-ups of radium dials in 1998 and 2003, and a clean-up by a toxic waste removal company in 2004.

Christmas Island in the Pacific, as it was then known, is now called Kiritimati, pronounced *Ki-ri-si-mas*, so for ease of reading, the Anglicised spelling will be used throughout.

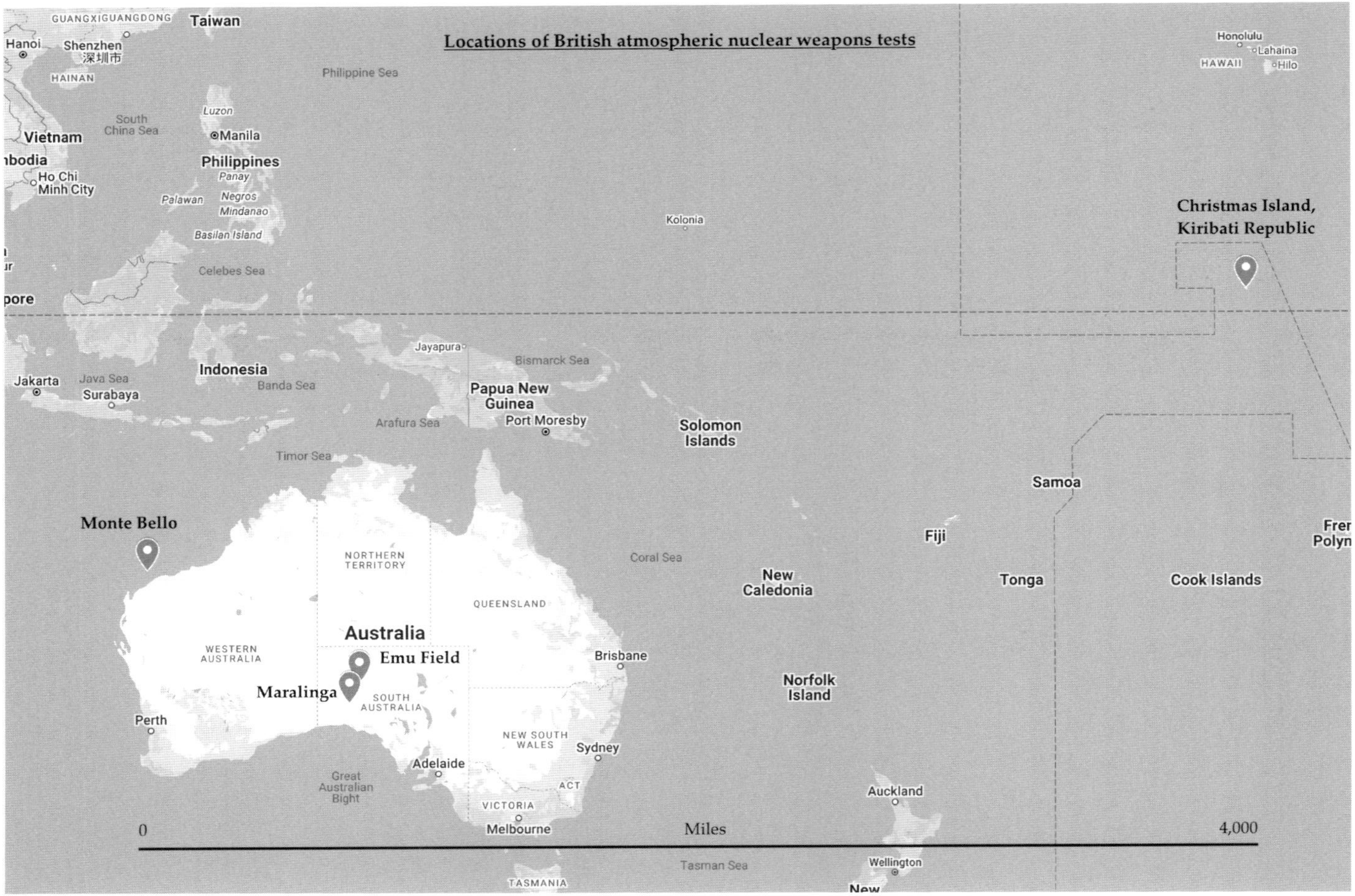

All map data ©2024 Google

Map of Christmas Island

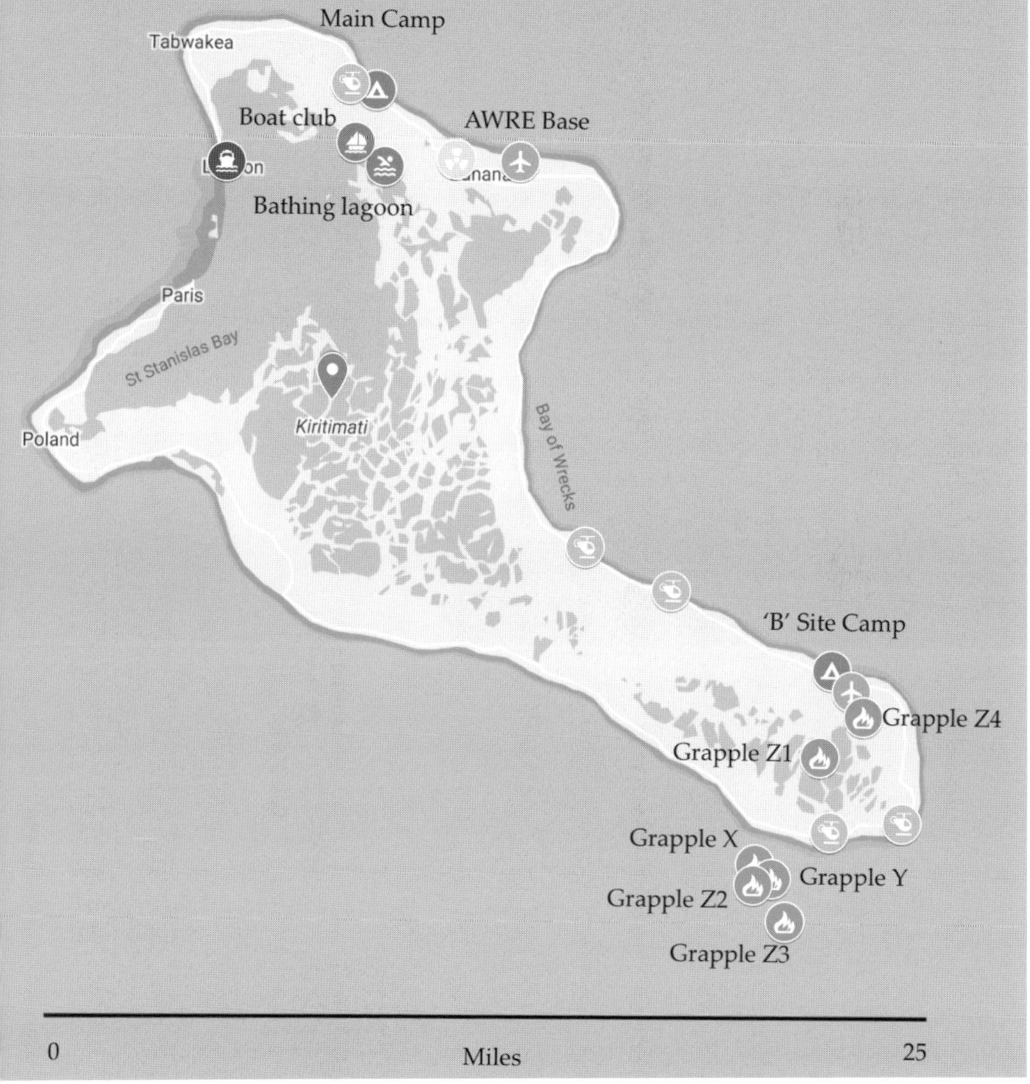

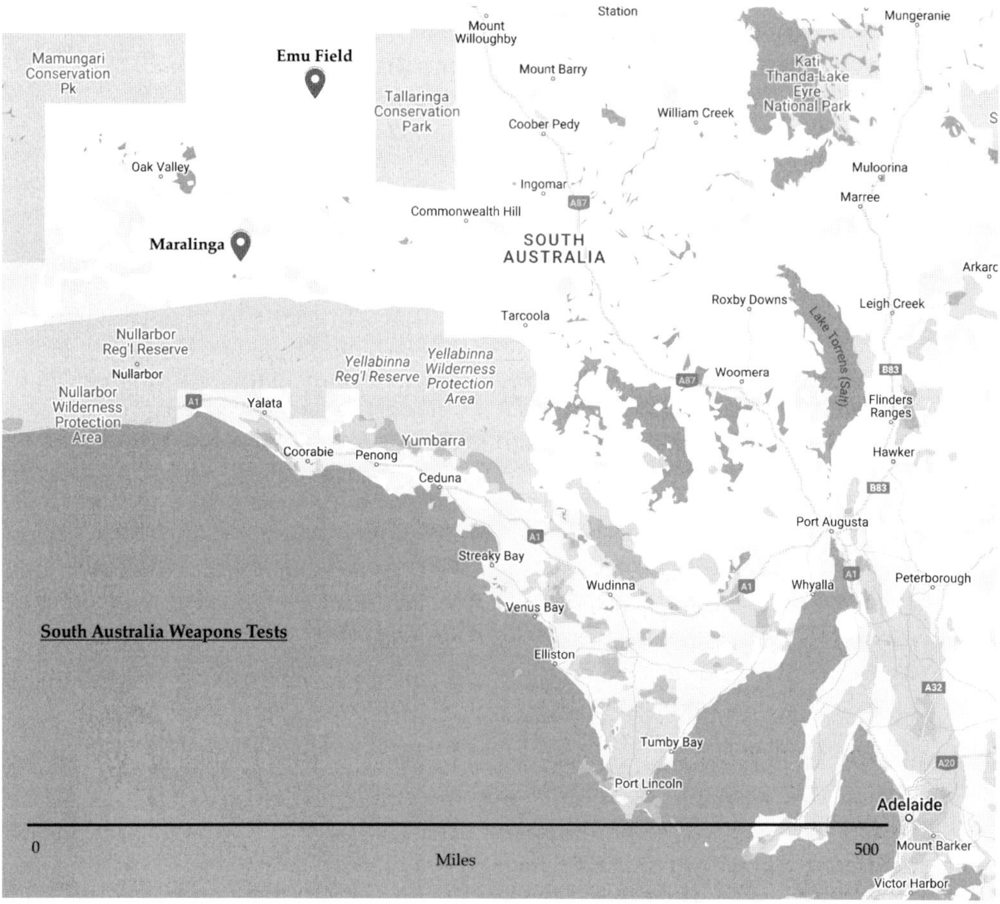

Emu Field
Maralinga
Mamungari Conservation Pk
Tallaringa Conservation Park
Mount Willoughby
Station
Mount Barry
Coober Pedy
William Creek
Kati Thanda-Lake Eyre National Park
Mungeranie
Oak Valley
Ingomar
Commonwealth Hill
Muloorina
Marree
SOUTH AUSTRALIA
Arkarc
Roxby Downs
Leigh Creek
Tarcoola
Lake Torrens (Salt)
Nullarbor Reg'l Reserve
Nullarbor
Nullarbor Wilderness Protection Area
Yellabinna Reg'l Reserve
Yellabinna Wilderness Protection Area
Woomera
Flinders Ranges
Yalata
Yumbarra
Coorabie
Penong
Ceduna
Hawker
Port Augusta
Streaky Bay
Wudinna
Whyalla
Peterborough
Venus Bay
South Australia Weapons Tests
Elliston
Tumby Bay
Port Lincoln
Adelaide
Mount Barker
Victor Harbor
0
500
Miles

INTRODUCTION

ONCE UPON A time, it was necessary to have a weapon that could unleash hell.

The Second World War was humanity's deadliest conflict, and both Nazis and Allies raced for the ultimate advantage. The US got there first, dropping atomic bombs on the Japanese cities of Hiroshima and Nagasaki in August 1945.

An account of the attack in *LIFE* magazine read: 'People's bodies were terribly squeezed, then their internal organs ruptured. Then the blast blew the broken bodies at 500 to 1,000 miles per hour through the flaming, rubble-filled air. Practically everybody within a radius of 6,500 feet was killed or seriously injured and all buildings crushed or disembowelled.'

A week after the second strike, it reported: 'When the [Nagasaki] bomb went off, a flier on another mission 250 miles away saw a huge ball of fiery yellow erupt. Others, nearer at hand, saw a big mushroom of dust and smoke billow darkly up to 20,000 feet, and then the same detached floating head as at Hiroshima. Twelve hours later Nagasaki was a mass of flame, palled by acrid smoke, its pyre still visible to pilots 200 miles away. The bombers reported that black smoke had shot up like a tremendous, ugly waterspout. With grim satisfaction, [physicists] declared that the "improved" second atomic bomb had already made the first one obsolete.'

A month later, photographer Bernard Hoffman wrote a memo to his picture editor describing what he had seen at Ground Zero. 'We saw Hiroshima today or what little is left of it,' he said. 'We were so shocked with what we saw that most of us felt like weeping... we were revolted by this new and terrible form of destruction. Compared to Hiroshima, Berlin, Hamburg and Cologne are practically untouched... The sickly sweet smell of death is everywhere.'[1]

Within a fortnight, Japanese doctors reported those killed by the blast had died instantly, but many survivors with small burns began falling ill. Their appetite failed, their hair fell out, and they began to bleed from the gums. They developed temperatures of 104, vomited blood, and died. Post mortem, they were found to have lost 86 percent of their white blood corpuscles.

The raids killed 200,000 people. But after a war that killed 56million in 30 countries, such a weapon seemed the only guarantee of survival. On August 15, Japan surrendered, and two years later, a new constitution written by the US was signed into force. It promised democracy and a pledge never to go to war again.

But others had seen the power the US could wield, and feared a new world order in which they held supremacy over every living thing. Throughout the 1930s and

1940s, geneticists had reported mutations in exposed plants and animals, sterility, and effects on their offspring. All nations with nuclear ambitions were aware of it, and it was these horrors that attracted them. One bomb could flatten a city, then deliver disease and deformity for decades, and render a massive area toxic for millennia. It was what made nuclear weapons terrifying, yet a must-have.

Refugees from the Nazis had been recruited into the Manhattan Project, the US-led programme that produced the first nuclear weapons. They now had to decide whether to return home, or stay and build more. Some were Jewish, others Communists, some so sickened by the war they'd survived they considered it a calling to end all future ones. This diaspora spread across the globe, to the Soviet Union, Europe and US, while others joined academic institutions and shared their knowledge. Governments worldwide accepted them, and happily put them to work. The Cold War had begun.

Building nuclear bombs is a dirty business. Britain was one of many countries to try, and one of the few to succeed. A small nation, wracked by poverty and rationing, defied American anger, Soviet spies and public outrage to create a bomb 150 times more powerful than those dropped on Japan. But victory came at a price. More than 22,000 servicemen and scientists took part in bomb trials in Australia and the Pacific, and just a handful are still alive. Their families report aggressive cancers, rare medical problems, high rates of miscarriage, and deformities, disability and death for their children – and grandchildren.

The same story is told by survivors on four continents of nuclear tests conducted by the US, China, Soviet Union, and France, as well as veterans of Commonwealth nations seconded to the British programme. The same patterns of disease and sterility are seen in civilian survivors: by 'Downwinders' in Nevada, by islanders in the Pacific, and by indigenous people whose homelands became laboratories for radiation experiments that would never be conducted today.

Much has changed since then. What was not only acceptable, but necessary, when memories of war were fresh and superpowers racing for dominance, now seems theoretical. But the fallout lingers. A black cloud still hangs over the 2,000 UK survivors and their 155,000 descendants. They mistrust governments, fear pregnancy, and labour under the belief they were used as guinea pigs.

The psychological impact is immense. Families report suicides, marriage breakdowns, alcoholism and mental illness. For decades the British state has fought its veterans, demanding they produce proof it is not in their power to provide. Successive governments have argued every war pension, rejected every legal action, and denied the existence of documentary evidence only for it later to be found in the archives. Scientific proof of genetic damage remains elusive, and is unlikely to be found in the veterans' lifetimes. Other nations have decided the risks and evidence are enough to merit apologies and compensation. In Britain alone, the government has refused to admit fault. Every change in that official stance has been the result of new evidence uncovered by the survivors or their families, and given the oxygen of publicity.

Whether the troops were on National Service, or had signed on for a career, they had no choice about nuclear weapons. Refusing to take part would have led to a jail cell, and allegations of treachery.

Most have long suspected, but never been able to prove, that they were as much a part of the experiment as the bombs themselves. In a different time, with different fears, it may have been necessary. But the first duty of a country to those who keep it safe is to protect them in return. Instead, Britain denied, dismissed, and abandoned its veterans and their families to decades of doubt, fear, and sickness. They were in every sense exposed: to radiation, to the elements, to the bone.

RADIATION

A BLINDING FLASH, an eerie silence, then the sky cracked. The sound reached those watching at the same time as the blast wave – a scorching 600mph wind, carrying with it the long, grumbling roar of the worst weapon known to man.

At 7.59am local time on October 3, 1952, Britain detonated its first nuclear bomb. Journalists who had been given prime viewing spots telegraphed awe-struck accounts to newspapers worldwide. Prime Minister Winston Churchill was jubilant, the scientists bursting with pride. But on a small island off the coast of Australia where the explosion had taken place, the fallout of Operation Hurricane was just beginning.

'We had no protective clothing. You wore shorts and sandals and if you remembered to grab your bush hat on the way, that was all you had,' said Derek Hickman, a Royal Engineer who was there aboard guard ship HMS *Zeebrugge*.

'They ordered us to muster on deck and turn our backs. We put our hands over our eyes and they counted down over the tannoy. There was a sharp flash, and I could see the bones in my hands like an X-ray. Then the sound and the wind, and they told us to turn and face it. The bomb was in the hull of a 1,400-ton warship and all that was left of her were a few fist-sized pieces of metal that fell like rain, and the shape of the frigate scorched on the sea bed.'

Britain had become the third nuclear power on Earth, but her audacious bid to build 'The Bomb' could not have been more serious. Fission – the ability to split an atom, releasing incredible power – was discovered in 1938. A year later scientists including Albert Einstein wrote to US President Teddy Roosevelt begging him to build a bomb before the Nazis did.[1]

Plutonium was discovered in 1940, and a viable weapon was dropped on Japan just 1,700 days later. In 1941 Britain decided that for its own safety and future influence in the world it must have its own atomic power, and wartime leader Winston Churchill set up a secret weapons programme codenamed Tube Alloys.

In charge was a mathematical physicist called Dr William Penney. The Gibraltar-born son of an army officer, he had spent the war working with the navy investigating shock waves, and designed mobile breakwaters known as Mulberry harbours to help with the D-Day landings. Just as the Allied invasion of France in 1944 was imminent, he was made head of a British delegation to the Manhattan Project - the secret US atomic research program based at Los Alamos in New Mexico. They would pool their knowledge to develop a weapon powerful enough to bring peace.

William Penney, in the back seat with his arm resting on the side of a Ford Deluxe convertible, pictured in the 1940s with other Los Alamos National Laboratory employees Beatrice Langer, Emil Konopinski, and Lawrence Langer, while working at Los Alamos. (National Security Research Center, via Wikimedia Commons)

Penney's tasks included calculating at what height to detonate the device, and predicting damage. He observed tests, and debated targets. Hiroshima was on a flat, wide river delta, with nothing to stop the blast's spread. Nagasaki - which replaced Kyoto as a target when that city was decided to be too culturally important - had hills, which could bounce the explosion back and increase damage. Penney witnessed the second attack from a circling B-29, and entered the area on foot a week later.[2]

Britain expected the Americans to share their knowledge, and in the immediate aftermath, new Prime Minister Clement Attlee ordered the production of a plutonium stockpile. Penney travelled to the Marshall Islands to observe two more US tests. While there, he earned the admiration of his American counterparts when their measuring devices were destroyed by one of the blasts. During his tour of the Japanese ground zero, he had noticed paint tins battered out of shape by the shockwave, and after carefully noting their distance from the epicentre had taken them away for detailed research. Before the Pacific tests, he filled cans with water and placed them at varying distances away from the device. He must have seemed like a nutty professor to US experts confident in the latest tech - but when their devices crumpled in the pressure wave, Penney's cans could be studied instead to accurately measure it.[3] Surely, both sides benefited from co-operation.

It was not to be. Within a year of Hiroshima, the Soviets began their own atomic programme, and America legislated to keep its nuclear secrets to itself. Britain decided to go it alone. As Foreign Secretary Ernest Bevin told the Cabinet: 'We've got to have it, and it's got to have a bloody Union Jack on it.'

Penney returned home, and set up a secret laboratory to develop the device at Fort Halstead, a Victorian armaments store so sleepy the only staff was a caretaker. Three miles south of leafy Sevenoaks in Kent, it was the perfect place to work unnoticed. The official go-ahead was given in January 1947. But then it hit a snag.

A key component in the team was a theoretical physicist called Klaus Fuchs, a refugee who had fled the Nazis following the Reichstag Fire in 1933. As a member of the German Communist Party, his life was in danger. While his family remained behind to smuggle Jews to safety, Klaus continued his studies at universities in Bristol and Edinburgh. A bid to become a British citizen was frozen when war broke out, and in 1940 he was interned. He wrote academic papers during detention, and in 1941 was recruited to work on the atomic project. After finally being granted citizenship, in 1943 he was sent to the US, and ended up at Los Alamos with Penney.

There, he helped calculate how to implode the plutonium core, creating sufficient power to split atoms. He was present at discussions about more advanced, thermonuclear weapons, witnessed tests, and was considered so useful the Americans requested he stay on in 1946 for subsequent trials. After official co-operation between the two countries ceased, it is thought the British used him as a spy to pass information back to Fort Halstead. If the allegation is true, it would be highly embarrassing: not just to foreign relations, but because of what Fuchs did next.

He returned to the UK to take up a job at the Atomic Energy Research Establishment in Harwell, which required MI5 security clearance. He got it. Then in 1949, the Soviets exploded their first atomic weapon, codenamed First Lightning. It was suspiciously similar to the Nagasaki device in almost every way. It was discovered spies had passed its blueprints to the Russians. The USSR had enough expertise to make its own bomb, but the leaks from Los Alamos helped it get there in double-quick time.

Klaus Fuchs, a German Communist, was recruited to work on the British and American atom bomb projects. He passed on his knowledge to the Soviets, singlehandedly creating equality of arms in the Cold War. (Wikimedia Commons)

Britain and America were still co-operating in other ways, including a code-breaking project called Venona, aimed at cracking Soviet intelligence transmissions. Some of the first messages it uncovered referenced an agent known as REST and CHARLES, who had provided information 'of great value' from the Manhattan Project.[4] By late 1949 the likely candidates were narrowed down to two - Fuchs, and another German-born scientist. MI5 began tapping Fuchs' phone and post, but found nothing. He was befriended by a Special Branch officer called Jim Skardon, who found a pretext for a series of interviews that enabled him to build a bond with Fuchs. When finally confronted, Fuchs denied it all. But a month later, his conscience led him to confess.

Fuchs had been passing information to the USSR for years, having been recruited by the friend of a man he was interned with. The courier was an academic called Ursula Kuczynski - codenamed SONIA - who would meet him regularly in the leafy market town of Banbury, Oxfordshire, and pass his documents on to Moscow. Among other things, he told the Soviets how to process uranium and how much to use, something it had taken the Americans two years, and $400m, to discover.

In 1950, Fuchs pleaded guilty to breaking the Official Secrets Act. He was given a 90-minute trial and the maximum sentence of 14 years in jail. MI5 was humiliated, having granted him top security clearance four years earlier. The Americans, fearing their ally had been infiltrated by the Red Army, withdrew the last dregs of co-operation. They refused even to provide a testing site.

Penney visited Fuchs in jail in 1952, in the same year he was named as the director of the new Atomic Weapons Research Establishment (AWRE). Fuchs was freed after nine years, emigrated to Communist East Germany, and went on to help the Soviets and Chinese with their bomb programmes. Had he not done so, the Cold War might have ended in a fireball.

Penney's team was now under enormous pressure. Without American support, and uncertain of who could be trusted, Britain had to prove its mettle. They needed a wide-open area, with barely any population for 100 miles. Aware of the risks of fallout, they wanted somewhere with prevailing winds that would blow it out to sea, but away from shipping lanes. They needed campsites, laboratories, radio equipment and workshops.

The most-likely form of attack was feared to be a Soviet vessel sailed into a British port with an atomic device on board. Simulating such an explosion might provide information helpful to the Americans, and be a way to restore relations. Penney identified the Monte Bello archipelago, a chain of 174 small islands about 80 miles off the north west coast of Australia, as the best place. In September 1950, PM Attlee wrote to Australian counterpart Robert Menzies asking 'whether the Australian government would be prepared in principle to agree that the first United Kingdom atomic weapons should be tested in Australian territory'.[5] Menzies, who considered himself 'British to the bootstraps', agreed without consulting his Cabinet. In a personal message to Menzies, Attlee warned: 'The area is not likely to be entirely free from contamination for about three years and we would hope for continuing Australian help in investigating the decay of contamination. During this time the area will be unsafe for human occupation or even visits by e.g. pearl fishermen who, we understand, at present go there from time to time and suitable measures will need to be taken to keep them away.'

This was not all strictly true - everyone knew radiation could last significantly longer. Menzies nevertheless agreed, but made no final decision until after a general election which he narrowly won, without having to admit to an imminent atomic test. The plan was not derailed by biological surveys which found the islands had more than 400 types of plants, insects, and animals, including 27 previously unknown to science, nor by a subsequent election in which Attlee was replaced by Churchill. The new PM's scientific adviser Frederick Lindemann, 1st Viscount Cherwell, briefed him that without atomic firearms Britain risked becoming, in effect, a colony of the US.

He warned: 'If we are unable to make the bombs ourselves and have to rely on the American army for this vital weapon, we shall sink to the rank of a second-class nation, only permitted to supply auxiliary troops like the native levies who were allowed small arms but not artillery.'[6]

The 1951 defection of Foreign Office diplomats Guy Burgess and Donald Maclean, who had both been spying for the Soviets for years, cemented the need for secrecy. America would not renew relations if Britain gave too much away to the Australians. A memo from the Commonwealth Relations Office to the UK High Commissioner in Canberra stressed the point. The note, written by head of the defence department Edwin Sykes and dated March 5, 1952, said: 'We hope that, now that the announcement is over, there will be no further pressure from the Australian side for fuller details, or for information about the grounds on which this assurance was given... If therefore any request for further information on this subject is made to you, you should do your best to dissuade the Australians from pressing it.'[7] Unaware of how much it was being kept in the dark, the Australian government agreed the blast could take place, and 1,400 troops were mobilised.

On a hot June day in 1952, a 26-year-old called Elizabeth rode out of Buckingham Palace to take part in her first Trooping the Colour as Queen.

The new monarch was pictured sidesaddle on her chestnut horse, Winston, as she accepted the salute of her armed forces and roars of applause from her people. On the same day, with no fanfare, frigate HMS *Plym* slipped out of the Thames estuary on a secret mission that has kept Elizabeth's country safe ever since.

The first wave of vessels had left five months earlier, when George VI was still on the throne. The fleet included tank landing ships *Narvik*, *Zeebrugge* and *Tracker*, each with 17 smaller craft on board capable of reaching the shore. As they steamed south, sailors had no idea the king had died. It was only after they left the port of Fremantle, Western Australia, they were told they now served a Queen, and that their task was to detonate an atomic bomb.

They were followed that summer by *Plym* and the flagship, HMS *Campania*. By then the experiment was no longer secret, nor was it popular. Protestors tried to block the roads in Sheerness, Kent, to stop bomb components being loaded. The radioactive core, a ball of plutonium-239 that weighed 7kg and was just 5cm in diameter, was taken out by flying boat and installed on *Plym* shortly before detonation. WJ Moyce, the scientist in charge of its transportation, had orders that if the plane got into trouble he was to parachute out with it strapped to his chest. It was in an unsinkable cork box, with a bag of dye that would react with sea water, so that it could be found by spotter planes even if he drowned.

James Stephenson remembered being given an unexplained posting by his company officer. 'We went for training and they started weeding us out, removing lads they thought were Communist sympathisers or not up to it. Nobody told us what it was about. One day at sea, when we couldn't tell anyone, they called us up on deck and told us we were going to be taking part in a nuclear explosion. We didn't mind – I had friends who were fighting in Korea and I was just glad not to be with them.'

Eric Waterfield was an 18-year-old regular with 17th Field Regiment, Royal Engineers, likewise handpicked for 'special duty'. After being mustered in

HMS *Plym*, a frigate at the end of her useful life, was used to simulate what would happen to a warship in Britain's first atomic bomb test in October, 1952. (Picture taken in 1943 by Royal Navy photographer and part of the Imperial War Museum Collection, via Wikimedia Commons)

Abergavenny with '100 or so other baffled sappers', they all took ship in Portsmouth for who-knew-where. Eric said: 'When they finally told us we were going to Monte Bello, we thought "where the f*** is that?" We thought we were going to visit another paradise island, but it turned out to be a disappointment. There was spinifex grass that was like steel wire and mangrove trees that at first glance looked like plum trees growing in a mist, except the things that looked like plums from a distance on a closer look turned out to be huge spiders, and the mist was spider webs.'

The British troops were joined by 11 Australian navy vessels, and preparations began. Engineers went ashore to build a control room, laboratories, landing stages, bunkers, and install generators. Electric cables were laid on the bed of the lagoon, and steel towers built on shore for sensors and cameras. The *Plym* was anchored to the seabed with massive blocks of concrete. Declassified documents claim every man had a 'dose badge', a piece of photographic film worn on the uniform to record radiation. Later research found they were given to 96% of those present.[8]

Early in the morning, and three days late due to bad weather, the plutonium was placed inside the bomb deep in the ship's hull, below the waterline, and the fleet sailed a short distance away. On board the *Plym*, a skeleton crew of six sailors washed their mess pans out and departed. Royal Navy mechanic Henry Carter was the last to leave, rowing a small boat carrying two scientists who had primed the weapon. Thirty years later, he recalled taking them to a spot already picked out, behind a nearby island under an overhanging bit of rock.

Henry recalled: 'A black heavy canvas tarpaulin was pulled over the boat so we were now in darkness. We all then draped jungle green towels over our heads and I pressed the palms of my hands into my eye sockets. I was dressed in shorts and a pair of shoes. At zero there was a blinding electric blue light of an intensity I had not seen before or since. I pressed my hands harder to my eyes, then I realised I could see the bones of my hands. It seemed that this light was passing through the tarpaulin and towel for about two to twelve seconds and there seemed to be two surges and two detonations with a continued rumbling and boiling sensation. My body seemed first to be compressed, and then billowing like a balloon.'[9]

The countdown was broadcast over a tannoy by a balding Welsh physicist called Ieuan Maddock, known forever after as The Count of Monte Bello. Eric and the other sappers were ordered onto *Zeebrugge*'s deck. 'We were told to turn our backs to it. There was a countdown and a huge flash, then we were told to turn and look at this enormous, black cloud. We could see the ripple coming under the sea, and then the sound followed it. It was like being punched on both ears, so loud it was painful,' said Eric, who has suffered from tinnitus ever since.

Journalists watched from 55 miles away, and within hours news went around the world. The *West Australian* reported, 'on the tick of 8 o'clock… pressmen heard a report like a clap of thunder, followed by a prolonged rumble like that of a train going through a tunnel. The air and ground shocks were sufficiently intense to cause slight pain in the ears. The immediate flash resembled the top quarter of a setting sun. A dense and magnificently turbulent cloud almost immediately shot to a height of 2,000ft. At first deep pink, it quickly changed to mauve in the centre, with pink towards the outside and brilliantly white turbulent edges. Within two minutes the cloud, which was still like a giant cauliflower, was 10,000ft high. A small pure-white milling ball rested on top… one hour after the explosion the atomic cloud - the only cloud in a clear blue sky - was beginning to take elongated stratus formation as strong upper winds whipped it north along the coast. Within two hours the only evidence of the atomic explosion was a faint wisp of cloud a few miles above the horizon. Soon after, it disintegrated and disappeared.'[10]

In the UK, the *Daily Mirror* declared: 'Today Britain is Great Britain again. The orange-coloured flash which marked the explosion over the Monte Bello Islands after Britain's first atomic bomb, did more than signal the unleashing of a new terrifying weapon of war. It changed the world. It signalled the undisputed return of Britain to her historic position as one of the great world Powers. Today she stands alongside America and Russia in possessing not only the secret of the atomic weapon, but also the power to produce it. It may be that in knowledge she is ahead of both. Today Britain is again the greatest European Power. She can defend herself, she can defend others. Now Western Europe knows that without Britain she cannot survive. Britain's A-bomb also lessens Western Europe's need to rely on America's strength.'[11] Its front page carried a photograph of the firestorm which rose over the ocean, growing 12,000 feet tall in four minutes. The editorial trumpeted how Britain was now a force to be reckoned with, diplomatically, economically, and militarily.

A scene from a Ministry of Supply film showing a delighted William Penney and awestruck Rear Admiral Torlesse aboard HMS *Campania*, 20 miles away. The film shows them with their backs turned to the blast, before spinning around to watch the Hurricane cloud as it developed into a mushroom. (Wikimedia Commons)

A photograph of the Operation Hurricane blast, taken during an early stage in the explosion after the initial orange flash had been enveloped by the great uprush of water. Immense clouds of smoke, steam and spray have burst into the air, and the warship has been vaporised.

Immediately afterwards Eric was ordered onto the islands to collect monitoring equipment, and canned food that had been buried by scientists to see how radioactive it would become. A day later, some of his mates caught some huge crabs that had washed up, and were boiling them up in a pot on an open fire. Eric said: 'Along came an officer who kicked the can over, and said the crabs were obviously dying of radiation sickness.'

Twenty days afterwards, Churchill told the House of Commons: 'The weapon was exploded in the morning of 3rd October. Thousands of tons of water and of mud and rock from the sea bottom were thrown many thousands of feet into the air and a high tidal wave was caused. The effects of blast and radioactive contamination extended over a wide area and HMS *Plym* was vaporised except for some red hot fragments which were scattered over one of the islands and started fires in the dry vegetation.

'Very soon after the explosion two naval officers undertook the dangerous task of flying helicopters over the heavily contaminated lagoon where *Plym* had lain. This was in order to take samples of the water so that its radio-activity could be measured. After a longer interval, scientists and service personnel in protective

Daily Mirror

SAT OCT. 4 1952

1½d

FORWARD WITH THE PEOPLE

No. 15,208

Registered at G.P.O. as a Newspaper.

THIS BANG HAS CHANGED THE WORLD

By BILL GREIG

TODAY Britain is GREAT BRITAIN again.

The orange-coloured flash which marked the explosion over the Monte Bello Islands after Britain's first atomic bomb, did more than signal the unleashing of a new and terrifying weapon of war.

It changed the world. It signalled the undisputed return of Britain to her historic position as one of the great world Powers.

Today she stands alongside America and Russia in possessing not only the secret of the atomic weapon, but also the power to produce it.

It may be that in knowledge she is ahead of both.

Today Britain is again the greatest European Power. She can defend herself, she can defend others.

Now Western Europe knows that without Britain she cannot survive.

Britain's A-bomb also lessens Western Europe's need to rely on America's strength.

Equals Now

To the Commonwealth it means that the Mother Country is once more able to deter the enemies of her sister nations in any part of the world.

Only a matter of weeks ago America and Australia refused Britain the right to sit on the Council planning the defence of the Pacific.

Will they now change their minds?

Not even America can now afford to be without the scientific knowledge that is in our possession.

She has refused before to share her secrets with us—although Britons helped her to make the first atomic bomb—on the ground that we had not the power of production.

Now we stand with her as an equal, proving that a nation needs neither the manpower and natural resources of Russia or industrial power of America to become supreme.

For America, isolation in knowledge is over.

Add to that our ability—unsurpassed in this world—to strike and defend in the air, [illegible] the Farnborough air show proved.

Add to that the results of our research which have made us the only nation in the world able to export the commercial products of atomic power.

The world asks: Was it H-bomb?

PICTURED above is the deadly radio-active cloud that rose over the Monte Bello Islands, off North-West Australia, yesterday, after scientists had exploded Britain's first atomic bomb.

Three facts set scientists all over the world guessing. The facts were:

1—*The cloud never took the familiar mushroom shape of the American atomic explosions. Before a strong wind blew it away, it looked like a gigantic "Z."*

2—*Observers said that the British bomb appeared to unleash far greater destructive power than the American bombs.*

3—*The flash was a deep orange colour.*

And several scientists asked: "Was Britain testing a form of the hydrogen bomb?"

Professor V. H. Bailey, Professor of Experimental Physics at Sydney University, Australia, said: "I do not know, and I would not like to say, that it was a hydrogen bomb, but if the cloud was different then it is apparent that the British bomb is different."

Mr. William Mangini, Australian research physicist, who witnessed the explosion from the Australian coast, said: "This

Continued on Back Page

Hit-and-run gunman shot her twice

Mrs. Mabel Whitehead, 32 — shot twice by a mystery gunman.

'SHE THOUGHT HIS PISTOL WAS A TOY . . .'

A BEAUTIFUL, thirty-two-year-old woman lay critically ill in hospital last night as an army of detectives and special police patrols sought a hit-and-run gunman who fired two shots into her stomach from close range.

She is Mrs. Mabel Whitehead, wife of a Skipton, Yorks, sales manager. Her husband, John, told the *Daily Mirror* last night: "At the bedside this morning she told me that the man who fired the shots at her the night before was someone she had never seen in her life.

"I know of no enemies Mabel could have had, nor anyone with a serious grudge against her," he added.

His wife he said, told him that she turned into the back street near their home, in Ermysted-street, and saw the man pointing what she thought was a toy pistol.

"He fired a shot, but she didn't realise what was happening until he fired a second bullet," Mr Whitehead continued.

"The man turned and ran into another side street and she staggered about fifty yards to our house and thumped on the door. She collapsed in the doorway.

"It was not until I returned home that I found out she had been shot.

"The shooting is senseless. My wife and I were completely happy. We had no worries."

They have been married ten years.

Earlier in the evening, he said, he had taken his wife by car to her mother's, where she was going to do a few household jobs.

In hospital, Mrs. Whitehead was able to give the police a statement.

She was given blood transfusions and morphia.

Within hours of the blast, the *Daily Mirror* in Britain had declared Operation Hurricane was 'the blast that changed the world'. (Reproduced with the permission of Reach plc)

clothing entered the contaminated area to examine the effect and to recover records.' He knew, but did not reveal, that the yield of the weapon was 25 kilotons, making it slightly bigger than that which had destroyed Nagasaki.

He said: 'Technical descriptions of the performance of the bomb cannot, of course, be given. It may, however, be said that the weapon behaved exactly as expected… To give some idea of the character of the explosion perhaps I might say this: normal blood temperature is 98.6 degrees. Many of us go over 100 degrees.

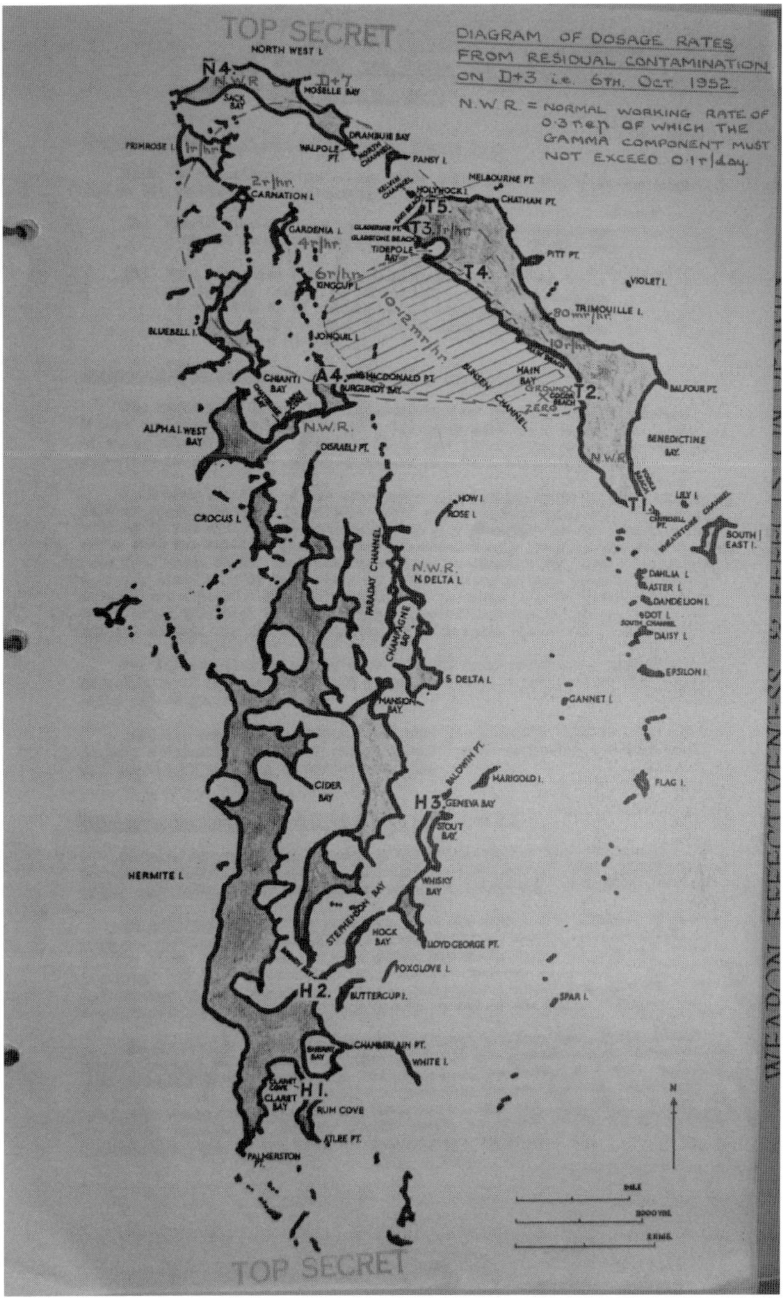

Official documents from the National Archives record the radiation readings, gathered by scientists and unprotected servicemen, after Operation Hurricane. The blast was centred on HMS *Plym*, moored near the spot marked ‘T2’.

When the flash first burst through the hull of *Plym* the temperature was nearly 1 million degrees. It was, of course, far higher at the point of the explosion. The explosion caused no casualties to the personnel of the expedition. No animals were used in the test. Apart from some local rats which were killed, no mammals were seen in the affected area and such birds as there were had mostly been frightened away by the earlier preparations.'

He predicted American co-operation would be re-established and added: 'All those concerned in the production of the first British atomic bomb are to be warmly congratulated on the successful outcome of an historic episode.'[12]

In a radio broadcast, Penney outlined the reasons why he thought it essential. 'The sight before our eyes was terrifying. A great greyish-black cloud being hurled thousands of feet into the air and increasing in size with astonishing rapidity. A great sandstorm suddenly sprang up over the islands,' he said. 'The energy and enthusiasm which have gone into the making of this new weapon stemmed essentially from the sober hope that it would bring us nearer the day when world war is universally seen to be unthinkable.'[13]

In Australia, RAF ground crew whose job was to wash down aircraft after they'd flown through the cloud fell sick, and low levels of radiation were detected all over the Australian mainland. But it wasn't the full story. Many years later, a 1954 document about scientific data collected at Hurricane was uncovered. Stamped 'SECRET - GUARD', it states: 'At various times, men entered the radioactive areas without respirators and it was possible to find traces of radioactivity in their urine within a few days.'[14]

It details where contamination was found and on how many men, even after showers. 'Neck 91, face 67, hair 67, knees and legs 54, hands 27, shoulders 10, arms 7, chest 3, feet 1,' it says. 'The location of the contamination varied with time. For example, about seven days after the burst, entry into areas of heavy contamination was possible and an increased number of contaminated knees and legs was caused by walking through knee-deep vegetation. Later the boats became contaminated and this was spread to the knees and legs by sitting on the decks.'

Although some men entering the danger zones were given respirators, oversuits, rubber boots and gloves, it was 'generally effective in preventing body contamination but where the suit was ruled or pressed in, e.g. at the knees or elbows or round the legs while pushing through the long spinifex grass, it penetrated the suit and contamination of the skin resulted… the hats provided were unsuccessful as they did not prevent dust-borne and airborne contamination from collecting on the exposed parts of the head and face.'

On at least one occasion, men were ordered into the fallout. The document says air samples were taken to measure how much radiation was in the air at head level, using mechanised sampling machines which sucked air across filter papers. It also reveals: 'In three cases it was arranged for dust to be stirred up by walking up and down in windward of the sampler while the sampling was in progress. A visible dust deposit was collected on the filter papers in consequence.'

It does not note whether those ordered to stir up the dust wore protective clothing; but in any event, such protection did not work.

'It was a plutonium bomb, and they're the dirtiest,' said Derek Hickman. 'After the tests, I married an Irish girl, she was magic. We started talking about families. A few years later I got sent to the doctor for something minor. I mentioned Monte Bello. He asked me if I was married and when I said "yes" he replied "my advice is never to have children". He wouldn't be drawn on why.' It was a warning Derek couldn't ignore.

'My wife wanted children and in the end, I walked away from the marriage. I thought if something went wrong I could never live with myself. I thought, to be fair to my wife, I should leave. I never told her why. A few years later she found out, and she said "I understand, but you should have talked to me". She never blamed me but it's the worst thing I've ever done in my life and I've regretted it ever since.' Derek sighed. 'She's passed away now, but she was a cracker.'

Seventy years later, he felt he and his comrades had been erased from history. 'We were kids. I was 19. We went off to a tropical island with our mates, happy as anything, and saw the most terrible thing anyone's ever seen. It's because of what we did that Britain has a nuclear deterrent. It's still keeping this country safe from Vladimir Putin and everything he's doing. Yet no-one knows about it. It's like we don't exist.'

Derek died in 2023 aged 90, acutely aware he was one of only a handful to make it that far. He retold his story whenever asked, and always spoke sadly of the wife he had left for the sake of their unborn children. 'There's very few of us left. It suited the government to never look at us again. But we know it was the dirtiest bomb in the world,' he said. 'But we were The Few - we were the only ones who served King, Queen, AND country, and there's not many who can say that.'

On his return home, Eric formed part of the honour guard for the Queen's coronation. Two years later, he had a large growth removed from his lung. The doctor told him it wasn't cancer, but didn't know what it was, and said: 'It's better in the bucket than in your chest.' He married, but soon after wife Barbara lost their first child, a son, to miscarriage. They emigrated to Canada in 1965 and had three daughters, all of whom had fertility issues. His three grandchildren also have health problems.

He said: 'Other countries recognised the risks, paid compensation, but in Britain we were told there was no problem at all. Well, we all knew that was BS.'

Churchill had given Parliament a version of events that was heavily edited, not just to appease the Americans or keep the scientists' secrets safe, but to minimise the risks run by servicemen. His papers from the period were closed for years. In February 1983, Margaret Thatcher told Parliament the records were not being released as expected under the 30-year rule as 'the material is still sensitive'.

Attempts to access them via the National Archives were refused, and early in 2022 a minister told Parliament that it had not been possible to find them. But a few months later they were quietly released. And the 87 pages of notes and memos show how Churchill took credit for the bomb, which he knew had contaminated the area, at the same time as erasing the troops who had made it possible.[15]

In a draft version of his speech, Churchill's red pencil - faded purple by the years - removed the line: 'Many of Her Majesty's subjects have played a worthy part in this remarkable experiment.' He deleted the fact he had already asked

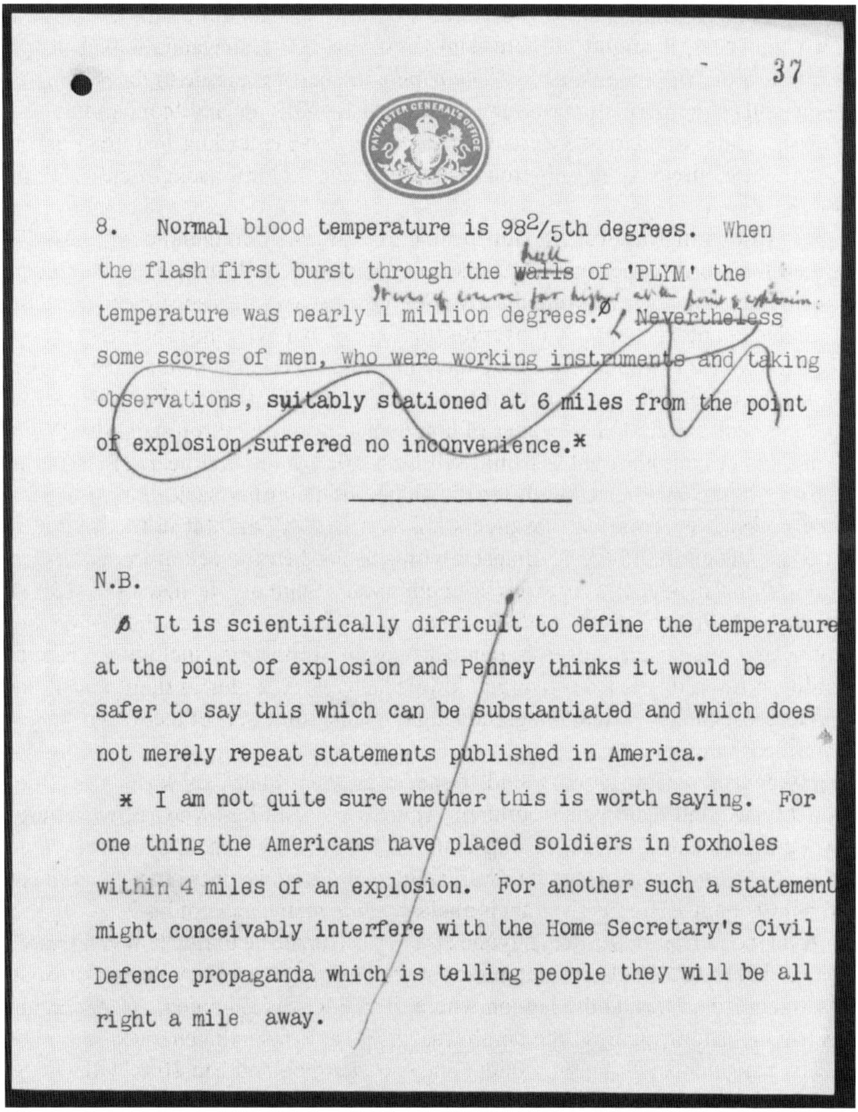

37

8. Normal blood temperature is 98$^{2}/_{5}$th degrees. When the flash first burst through the walls of 'PLYM' the temperature was nearly 1 million degrees.ø Nevertheless some scores of men, who were working instruments and taking observations, suitably stationed at 6 miles from the point of explosion, suffered no inconvenience.*

N.B.

ø It is scientifically difficult to define the temperature at the point of explosion and Penney thinks it would be safer to say this which can be substantiated and which does not merely repeat statements published in America.

* I am not quite sure whether this is worth saying. For one thing the Americans have placed soldiers in foxholes within 4 miles of an explosion. For another such a statement might conceivably interfere with the Home Secretary's Civil Defence propaganda which is telling people they will be all right a mile away.

A page from Winston Churchill's draft speech about Britain's first atomic blast shows his red pen removed claims that servicemen had 'suffered no inconvenience' from the radiation. (National Archives reference PREM11/563)

for Penney to be honoured by the Queen, which would have made the lack of recognition for others all the more obvious.

He removed a line claiming 'scores of men suitably stationed at six miles from the point of explosion suffered no inconvenience'. It would have exposed a lie being told to the public at the time, that nuclear weapons would not contaminate huge areas. In the notes, an official warns the PM: 'I am not quite sure whether this is worth saying. It might interfere with Civil Defence propaganda telling people they will be all right a mile away.' The papers include a message to the Admiralty, saying fires were burning across the islands for eight hours and 'contamination is intense' a mile from Ground Zero. One instrument was recording 100 Roentgen an hour - 57,000 times more radiation than the annual 'natural background' dose for the UK.

Yet Churchill was warned not to mention 'areas of contamination'. And he deleted the word 'experiment'. As the PM knew full well, men were being sent deep into the danger zone immediately after the blast to collect specimens and data. While the scientists at least had overalls, the servicemen wore just shorts and boots.

Royal Marine Kenneth Gower, then 19, was ordered to stand knee-deep in the sea to collect dead fish and turtles, wearing just a pair of shorts. Daughter Janice Parrott said: 'He had waves of horrendous vomiting throughout the 1960s, so bad the doctors thought it would kill him. He had his first heart attack in his 40s, and in the 1980s had such terrific skin lesions on his back, legs and arms. They called it eczema, but the ulcers left him with a large pit in the middle of his chest.' Then in 2000 he collapsed with another heart attack at the Cenotaph, while laying a poppy wreath on Remembrance Sunday. He had bypasses, an internal defibrillator, then bladder cancer and a cancerous growth on one eye. Janice and her sister suffer a range of health problems, including immune condition sarcoidosis, fibromyalgia, kidney infections, holes in their bones, and skin lesions. One of Kenneth's granddaughters had three sets of teeth, and needed painful extractions.

Janice said: 'Mum always said if she knew then what they know now, they probably wouldn't have had children. There were families who suffered more, and bore financial burdens, raising children alone and so on. Considering dad's exposure we were very lucky. He was proud of being a marine, but very unhappy that other countries recognised their nuclear test veterans and not his.'

Alan Ayres was closer than anyone else to the Hurricane blast. He was Penney's official driver. On the day of detonation it was Alan's job to ferry junior scientists to a sandy island beside the lagoon where the *Plym* was anchored, and where the electric timer for triggering the bomb was wired up to a small generator.

His daughter Shelley Ayres said: 'They set the time and had 20 minutes to get back in the Land Rover, across the island, and into a boat that would take them out to sea. Only the car got stuck in the sand, and they had to run for it. There was one scientist who wasn't going to make it, so dad put him on his back and carried him. They had to shelter behind a hill, called H1, underneath a piece of canvas. He always said he was right there, when the bomb went off, and just had to turn their backs and cover their eyes.'

When he came home, Alan became a bus driver, and met wife Pam, a 'clippy' who checked the tickets on his route. Shelley said: 'He always had terrible, shooting pains throughout his body ever since I can remember, without any explanation, and he'd always say it was something to do with the atomic bomb. We're so proud

Alan Ayres, pictured with his feet on the mudguard of his army truck, before he was sent to Operation Hurricane, where he would be closer than anyone else to the blast. (Courtesy of the Ayres family)

he was there, but we just want the recognition... so many people have no idea it happened. The government behaved as though it was shameful. I'm very proud my dad is still keeping this country safe today.'

James Stephenson was medically discharged after his return to Britain and married, fathering two children. Both were healthy but he said: 'I know people whose children were born with their organs outside their bodies. It wasn't everyone, but stories like that weren't unusual. I'd had my children but it made me worry about my two grandchildren before they were born, and thank God they're fine. I've got a respiratory problem and have to breathe with a nebuliser but I've been lucky really. When we left home King George VI was on the throne and when we came back it was his daughter. I don't think anyone really knew what they were doing, not us or the scientists. It was just a job we had to do.' James died in September 2021, aged 89, after a fall at home. He had suffered a lung condition for many years which he attributed to fallout, and needed antibiotics every day to keep his airways clear.

Kenneth McCormack had been a carefree, 21-year-old REME corporal when he was sent to Monte Bello. When the explosion happened, he was ordered to witness it from the deck of an observation ship anchored just 12 miles away.

Afterwards he returned home to Birmingham, married his sweetheart Muriel, and fathered seven children - five of whom had abnormalities. A daughter needed spectacles from 18 months, a son had a condition that made his eyelashes grow into his eyeballs, another girl was born with a hole in her eye and another son had severe breathing difficulties. But when his son Peter died in 1984 aged 28, of aggressive liver cancer, Kenneth decided there was a link with the tests.

Muriel, who also had one miscarriage and a stillbirth, said: 'Kenneth blamed himself, and said he had murdered Peter. He died himself a year later in 1985 of a heart attack. I am sure both their deaths were related to the radiation Kenneth absorbed. I just want the authorities to be honest with us. I have written to every Prime Minister in succession asking for them to recognise the families' problems… I will fight for the truth until I die.' She passed away in 2022, still fighting.

Penney was knighted, but despite Churchill's optimism American co-operation was not renewed. Less than four weeks later the US detonated a hydrogen bomb, a more advanced nuclear device that was 400 times more powerful, and Britain was back out in the cold.

The Monte Bello islands are now a wildlife park, but visitors are warned not to stay for more than an hour or take home the fragments of metal that can still be found on the beach - radioactive remnants of a long-forgotten hurricane.

Attlee had used sleight-of-hand accounting to secretly spend £100million on the atomic project, equivalent to £3.6billion today, before informing Parliament of his plans in 1948. The sums were so vast, and the potential gains so vital, that Britain was now committed to the nuclear arms race – and, despite the effects of crippling post-war austerity, it was imperative to keep going.

Penney got approval for a new series of experiments to refine bomb components. Although it was unpopulated, to hold them at Monte Bello would have demanded massive and sustained naval support over years. Instead, a new testing site was identified at an isolated patch of sandstone, 300 miles north west of the outback town of Woomera in South Australia.

The Woomera Rocket Area was created in 1947 to test British missiles, but it wasn't popular. Public meetings were held by campaigners including the Council

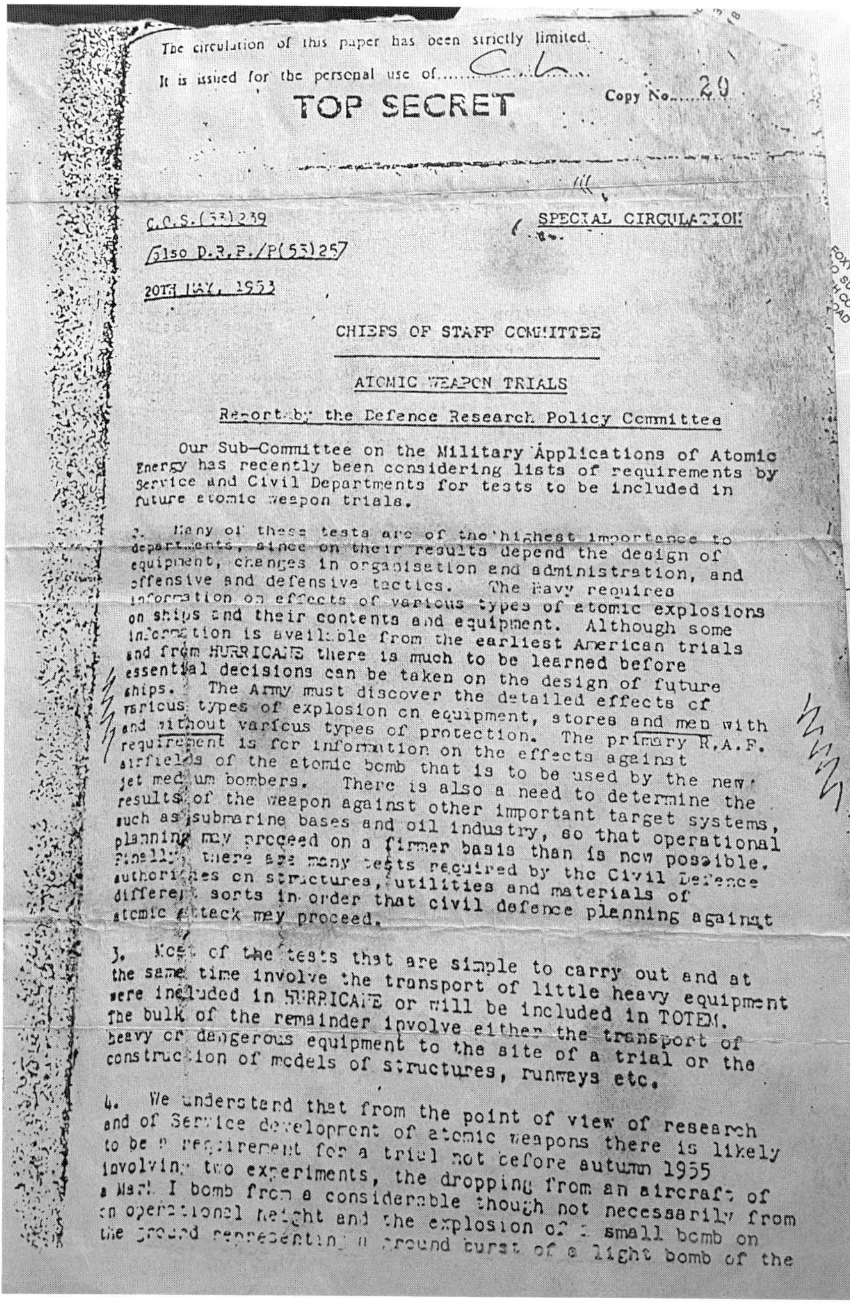

The circulation of this paper has been strictly limited.

It is issued for the personal use of ......C.L......

Copy No......20

TOP SECRET

C.O.S.(53)239

SPECIAL CIRCULATION

6150 D.R.P./P(53)257

20TH MAY, 1953

CHIEFS OF STAFF COMMITTEE

ATOMIC WEAPON TRIALS

Report by the Defence Research Policy Committee

Our Sub-Committee on the Military Applications of Atomic Energy has recently been considering lists of requirements by Service and Civil Departments for tests to be included in future atomic weapon trials.

2. Many of these tests are of the highest importance to departments, since on their results depend the design of equipment, changes in organisation and administration, and offensive and defensive tactics. The Navy requires information on effects of various types of atomic explosions on ships and their contents and equipment. Although some information is available from the earliest American trials and from HURRICANE there is much to be learned before essential decisions can be taken on the design of future ships. The Army must discover the detailed effects of various types of explosion on equipment, stores and men with and without various types of protection. The primary R.A.F. requirement is for information on the effects against airfields of the atomic bomb that is to be used by the new jet medium bombers. There is also a need to determine the results of the weapon against other important target systems, such as submarine bases and oil industry, so that operational planning may proceed on a firmer basis than is now possible. Finally, there are many tests required by the Civil Defence authorities on structures, utilities and materials of different sorts in order that civil defence planning against atomic attack may proceed.

3. Most of the tests that are simple to carry out and at the same time involve the transport of little heavy equipment were included in HURRICANE or will be included in TOTEM. The bulk of the remainder involve either the transport of heavy or dangerous equipment to the site of a trial or the construction of models of structures, runways etc.

4. We understand that from the point of view of research and of Service development of atomic weapons there is likely to be a requirement for a trial not before autumn 1955 involving two experiments, the dropping from an aircraft of a Mark I bomb from a considerable though not necessarily from an operational height and the explosion of a small bomb on the ground representing a ground burst of a light bomb of the

The UK Chiefs of Staff meeting in May 1953 heard that further atomic tests were needed to ‘discover the detailed effects of various types of explosion on equipment, stores and men, with and without various types of protection’.

for Aboriginal Rights, Quakers, Communists, and women's temperance unions, who wanted the area left alone. But there was a place known as Emu Field, 300 miles north west, yet close enough to be supported by the existing firing range. Australian officials said it was so isolated that 'the area is no longer used by Aborigines'. The Brits decided it was perfect.

In May 1953, an MoD report for the chiefs of staff detailed the need for more experimentation. 'These tests are of the highest importance to departments, since on their results depend the design of equipment, changes in organisation and administration, and offensive and defensive tactics. The Navy requires information on effects of various types of atomic explosions on ships and their contents and equipment... there is much to be learned before essential decisions can be taken on the design of future ships. The Army must discover the detailed effects of various types of explosion on equipment, stores and men, with and without various types of protection.'[16] Two years later, an RAF memo stated it expected to 'gain invaluable experience in handling the weapons and demonstrating at first hand the effects of nuclear explosions on personnel and equipment'.

In June 1953, Penney assured Australian observers there was no risk of contamination. They in turn told Menzies: 'The isolation of the trials precludes any possible damage to habitation or living beings… the time of firing will be chosen so that any risk to health due to radioactive contamination to our cities, or in fact of any human beings, is impossible.'[17] This assertion was utterly wrong. The 2,000 or so semi-nomadic tribespeople for whom the area was home tended to avoid those who came looking, but it was known they were there. As the tests approached, any who could be found were forcibly relocated, on foot or by truck, to Christian missions which were a long way from their traditional water sources, hunting grounds, and normal lifestyle.

Not only was little known about the people who lived there, a branch of the Anangu tribe known as the Tjarutja, but no real effort was made to find out. At the time, no census included indigenous people, who were often listed as livestock on farms. As far as white Australia was concerned, they would undergo a process of 'detribalisation' and adopt European ways, which would mean their reserves could be used for other things. As one official wrote in 1955, Australian policy was 'to assimilate the natives into the white man's way of living as quickly as possible'.

But these inconvenient people still had the potential to disrupt a delicately-timed programme through accident or mischief, so an order went out to evacuate any Aboriginal who lingered. The British ordered patrols before the tests, and sent warnings to far-flung cattle stations warning them not to let anyone come too close. But the range was 38,000 square miles, and the task next to impossible. It was not enough.

Royal Engineers prepared the testing ground, with an airstrip built to bring in thousands of tons of equipment, and a village created for servicemen and scientists to live in. The stage was set for Operation Totem in October 1953. The Soviets exploded their first hydrogen bomb that August, and Britain was falling behind in the race.

But first, and without informing the Australians, in the month beforehand Penney's men held a series of tests codenamed Kittens, testing trigger devices used to start the chain reaction in the heart of the bomb. Five small explosions were carried out near Emu Field, which were later found to have peppered an area about 400 yards wide with small amounts of heavy metal beryllium and radioactive polonium-210, both highly toxic.

For the main event, top brass ordered up a special piece of equipment. A 40-ton Centurion tank would be parked 500 yards from the Totem blast to see how standard-issue military kit could be affected by the new weapon. Items used for such destructive experiments were usually defective, but this tank had rolled off the production line only two years earlier and had just 500 miles on the clock. Its use was symbolic of how afraid governments were that conventional hardware would be rendered useless by a nuclear-armed enemy.

This one - a Mark 3 with the serial number 169041 - was transferred with its crew from the training facility in Victoria by train, then loaded onto a truck and trailer to be carried into the interior. At places, it had to be unloaded and used to haul the truck, before being reloaded and continuing as before. The final 170 miles were so inhospitable the tank had to go on alone.

Once at the range, it was filled with sensors, standard ammunition, and a fake 'crew' of shop mannequins whose later 'injuries' could be studied. Shortly before the first test on October 15, the Centurion's engine was fired up and the hatches closed before the scientists retreated to a safe distance. Operation Totem consisted

What became known as 'The Atomic Tank', pictured on its way across the Outback to the remote firing range. The area was so inhospitable that in places it had to be offloaded from its transporter, and used to haul the truck out of trouble, before the convoy could proceed. (Courtesy of Roy Vincent)

of two explosions of plutonium weapons at the top of 100ft towers, two weeks apart, with a yield of 10 and 8 kilotons respectively.

After the first, the *South Australia News* reported: 'The British atomic bomb, the first in history to be exploded over land, today blew the top off the down in a South Australian desert with a crack that, it was said, could be heard 200 miles away and with overtones that echoed across the world... a device of unprecedented dimensions, supported by undescribed means, at an undefined height over the clay pan, thudded into all corners of the morning with a noise that took 57 seconds to reach us.'[18]

When the tank was examined, it had shifted 5ft backwards. The hatches were blown open, the antennae ripped off, its paint sand-blasted. The armoured plating, crumpled past any use, was found 200 yards away. The engine had also shut down - but only because it ran out of fuel. The crew, it was thought, would have had their insides turned to jelly by the shockwave.

Officials were delighted. Despite being unable to protect its human inhabitants, the tank, which was more expensive, still functioned. Three days after the test, its purpose fulfilled and without any decontamination, the crew was ordered to drive it back to Woomera. Again, the tank had to tow its own transport part of the way, and the trailer had several blown tyres. After it returned to base, the crew - who wore no protective clothing - were ordered to park it as far away as possible before it was washed down.

But that wasn't the end of the Atomic Tank's story. With a new engine it was used as a training vehicle, had several upgrades, and was sent to Vietnam in 1968. A year later it was hit by a rocket-propelled grenade during a firefight with the Vietcong. After the war it returned to Australia, and was used in military parades due to its 'indestructible' reputation. It is now regarded as military memorabilia, and parked up at the regimental base.

But in 1990, the *Geelong Advertiser* reported 12 out of 16 men who operated the tank after the tests had died of cancer. Former Warrant Officer Class One Bob Thompson, by then a secondary school teacher, told the newspaper that on a hot day soldiers would rest on the turret, and later developed severe boils on their backs and legs. Mr Thompson, then aged 59 and diagnosed with prostate cancer, said their commanding officer warned them not to touch the tank but, without any explanation, the men ignored it. 'We needed it for parts,' he said.

Totem was also the first time that an air crew was ordered to fly through the mushroom cloud. After Totem 1, Wing Commander Geoffrey Dhenin took a high-altitude Canberra bomber to 30,000 ft to take a 'cut' or fly 'through the fringe of the cloud' six minutes after detonation. Codenamed Operation Hot Box, the aim was to take samples from within the cloud to enable scientists to calculate its true power and formation. There was a second, deeper 'run through the middle' after nine minutes, and further runs to help estimate the height of the cloud.[19]

The bomber was fitted with filters on its wing tips. Dhenin was a medical doctor, seconded from Bomber Command along with a highly-decorated colleague as navigator, and Group Captain DA Wilson joined them as radiological observer.

His report for AWRE states: 'The dose rate meters on the aircraft started to register some 500 yards away from the cloud and whilst in the cloud, showed a peak of approximately 2,000 Roentgen an hour... the cloud was very thick and gave the appearance of a dense London fog. Turbulence was considerable, but abrupt and short in character.'

He added further cuts were made three times at varying heights, with dosage rates of 300 and 400 Roentgen an hour. After landing and jettisoning the wing tip filters, the dose rate in the cockpit dropped a bit. As for the crew, he said: 'After being monitored, they went to breakfast.' The plane on the other hand was thoroughly decontaminated, and left to stand on a remote part of the airstrip for 48 hours. It took another 48 hours of 'heavy work' to reduce its contamination enough to be at operational levels.

Wilson said there were discrepancies between the dosimeter badges they wore, with some showing half the rate of others, and some reading 'off-scale'. The report said they did not understand why, and the crew had asked to fly a mission through the second cloud as well. But Penney decided the risk wasn't worth it. He told Dhenin: 'Go home, boy. You have done enough. I cannot authorise such a thing a second time.'[20]

The Hot Box report said the crew had blood and urine counts taken, and these 'failed to show the presence of any ingested fission products'. It concluded the sampling was 'without undue hazard to personnel'. Geoff Dhenin got the Air Force Cross, and enjoyed a long life. He retired in 1978 as Air Marshal, and was knighted a year later. According to his family, he always said a little radiation was good for you.

Despite being smaller than Hurricane, the Totem blasts were not a failure. They had finessed the amount of plutonium needed to produce a particular yield. But it is one of the deadliest natural substances on Earth, capable of lodging in the human body, and with a half-life decay of up to 24,100 years. A lack of wind meant Totem I's cloud hung in the sky 24 hours later. Fallout on the ground was the size, shape and colour of lead shot - small, black balls of one of the most toxic elements known to man, scattered across more than 300 miles of Outback.

It was also reported that a black mist had been seen rolling across the plateau. Similar ground-level clouds were documented at US tests in Nevada, but the authorities insisted it was impossible here. Even decades later in 1980, Professor Sir Ernest Titterton, a British physicist who became chair of the Australian government's Atomic Weapons Test Safety Committee, said in a radio broadcast aimed at quelling continuing rumours: 'No such thing can possibly occur. I don't know of any black mists.'

Yami Lester was only 10 years old in 1953, and living with his family in the Yankunytjatjara tribal lands. When Totem I exploded, they were working as farm hands and just waking up, 107 miles away at the Wallatinna cattle station. In his autobiography, Yami wrote: 'I remember the noise. It was a strange noise, not loud, not like anything I'd ever heard before. The earth shook at the same time; we could feel the whole place move.'

Within hours a large black cloud appeared low on the horizon. In his book *Maralinga*, journalist Frank Walker interviewed survivors, describing it as 'black, greasy-looking, and the sun shining on it glinted back, like it was reflecting off something solid. It was more like a black mist or dust storm. But there was no wind'. As the mist moved across the camp, terrified adults dug holes to shelter in with their children. They thought it was a mamu, or evil spirit. It left behind it a black, sticky substance on leaves, huts and the ground.

Yami remembered: 'A few hours later we all got crook, every one of us. We were all vomiting; we had diarrhoea, skin rashes and sore eyes. I had really sore eyes. They were so sore I couldn't open them for two or three weeks. Some of the older people, they died. They were too weak to survive all of the sickness. The closest clinic was 400 miles away.' He said that his uncle was among those who died, with sores all over his body full of pus. Others bled from the nose and mouth, with green vomit, green faeces. Yami lost the sight in one eye, and by 1957 was completely blind. The tribe almost starved as people were too sick to go hunting, and some died of thirst because they were too weak to scrape away the black scum that had settled onto waterholes. The frailest died within days, others lingered for a year. No doctor came to the camp, and the Aboriginal custom of not discussing the dead meant the mist passed into oral folklore.

In 1980, by then a tribal elder, Yami heard Professor Titterton on the radio talking about 'the Aboriginal people we looked after'. Enraged, he rang the *Adelaide Advertiser* and told his story. The front page, headlined 'A Black Mist That Brought Death' caused a national stir. Others came forward, indigenous people and white settlers, to confirm the tale. There were claims of Aboriginal mass graves, and white men taking away the corpses of those who had died. It was discovered that cancer rates in tribal communities had shot up from just one case in every five years, to ten.

But the British denied even the possibility it was true. An internal MoD memo discussing Yami's story said: 'The 1953 shots were tower shots and there is no way that any phenomenon corresponding to this description could have been experienced at a distance of 170 km NE of Emu Field.'[21]

In 1981, Lalli Lennon came forward. She had been a young mum-of-three gathering opals about 112 miles north east of the blast, at Mintabie. One of the government's native patrol officers had warned her to stay away from Emu Field. She told a TV crew they were looking for opal to sell, and had heard on the radio a bomb was due to go off early in the morning. Then she heard 'a big bang like a thunder storm, then it got louder, then it just vibrated'.

She said: 'Just as we were getting ready for breakfast, this smoke, you know, you could see it between the trees coming through, it went right through, over us. Could smell the gunpowder smell.' She rushed into a tent with her children, and when they came out, 'on the tent it had this grey, blacky sort of dust'. That night her children fell ill with fevers and headaches, and the whole family suffered diarrhoea and vomiting. Within two weeks Lalli and her baby son, Bruce, developed burns so

A concrete obelisk today marks the Totem I test site, warning visitors that radiation hazards still exist near where the 10 kiloton device was exploded in 1953. (Wikimedia Commons)

bad they 'felt like being rolled in a fire'. Even 30 years later they both still suffered from skin lesions erupting in cycles, leaving enormous scabs. Doctors were unable to diagnose it for decades, before deciding it was psoriasis.[22]

A few years later, Almerta Lander testified that she had been about 300 miles north of the blast zone, near the isolated Welbourn Hill cattle station. There were no clouds in the sky, so she took note of what looked like a 'black banner with tendrils coming off it'. She went on: 'The grey dust that fell was very sticky; very, very fine, soft and sticky. It was so fine that when you brushed it away it just rose up and settled down again. It was a beige-brown colour. We had to use a wet cloth to get it out of the caravan but it just smeared everywhere.' She cleaned off her pots and pans, and used them to cook. Her son was later diagnosed with lung disease, and Almerta developed a severe rash that never went away.[23] Similar burning skin problems were noted following black rain which fell on Hiroshima after 1945, on sheep following blasts in Utah in 1953, on Japanese fishermen after US tests in the Pacific in 1954, and on islanders in a report dated 1957.

Settlements in the cloud's path were found to have high levels of radioactivity. Estimates of the number of Aboriginal people affected varied, from 45 to 1,800, with perhaps 50 deaths in the aftermath and dozens more in the decades that followed. But as no medical or census records were kept of indigenous people in the 1950s, it was impossible to prove. And when PM Menzies was asked about whether checks would be made on local human, animal and plant life, he relied on assertions he'd received from the British, and accused critics of harming Australia.

'It has been stated most authoritatively that no conceivable injury to life, limb or property could emerge from the tests,' he told parliament. 'The tests are conducted in the vast spaces in the centre of Australia and, if it is to be said however groundlessly, that there are risks, what will be said in other countries? Are we then to reach the position in which we shall not conduct these experiments? Believe me, the enemy will conduct them… [we would be] contracting out of the common defence of the free world. No risk is involved in this matter. The greatest risk is that we may become inferior in potential military strength to the potential of the enemy.'[24]

Totem II exploded 12 days after the first. Frank Walker reported that the wind changed direction, sending the subsequent fallout cloud over the camp where the military personnel and scientists were living and working. 'The alarm went off and everybody dropped what they were doing, jumped into vehicles and got the hell out of there,' he wrote. Three years later, members of the Australian military returned and found tools, food, tents and tables lying where they were left.

Those tasked with keeping an eye on Aboriginal welfare began raising concerns about roads being laid in their territory, and damage to traditional lifestyles. Native patrol officer Walter MacDougall threatened to go public with allegations that their human rights were being infringed. He wrote about meeting one elderly tribesman, saying: 'The old man enquired for what purpose the strangers had visited his country, and expressed anxiety when told that they intended establishing a post. The word used means fear - I have not a word for troubled or anxious, but I gather that he was more than just afraid. That he was worried or concerned.'

Australia's chief scientist on the project, Alan Butement, wrote a memo about MacDougall to a colleague, summing up the official attitude: 'He is apparently placing the affairs of a handful of natives above those of the British Commonwealth of Nations.'[25]

Back in Britain, there was growing disquiet the all-powerful weapon they now had was not enough. The Soviets were racing ahead, and the Americans detonated the world's biggest hydrogen bomb in 1954. At 15 megatons, this 'super bomb' was 1,000 times more powerful than the combined yield of the Totem blasts. Churchill realised that his own atomic weapon was a pea-shooter by comparison.

He told the House of Commons: 'The advance of the hydrogen bomb has fundamentally altered the entire problems of defence, and considerations founded even upon the atom bomb have become obsolescent, almost old-fashioned.'[26] By June of that year, he had authorised the production of a British H-bomb. At the same time, he made a formal request to the Australians for a permanent testing ground. They had considered Indian Ocean atolls, the Falkland Islands, the British colony of Somaliland, the Bahamas, even a small island off the coast of Tasmania. Each was too populated, too American, or too windy. Then word came back of a place that might suit, 130 miles to the south of Emu Field, which had never seen white men before. The scrubland had low hills to protect observers, a nearby railway line, and fresh water from a borehole. Penney flew in for a secret inspection, and asked if there were Aboriginal people living nearby. Aside from an ancient ceremonial ground that was possibly neolithic, he was assured the local tracks were used only by 'one or two elderly blacks, and then on rare occasions'.[27]

Penney pronounced it 'very satisfactory'. Now he had a suitable site on which to hold hundreds of highly-toxic experiments as his team narrowed in on a weapon that would finally match the might of the Americans. To show the gravity of their intention, the area was given a name, from an Aboriginal dialect. It was called Maralinga - 'field of thunder'.

In 1954, the two nations signed a 10-year lease agreement, stating the British had responsibility for protecting people and property, while Australia would monitor fallout. Australian Minister of Supply Howard Beale said: 'England has the know-how; we have the open spaces, much technical skill and a great willingness to help the Motherland. Between us we should help to build the defences of the free world, and make historic advances in harnessing the forces of nature.'[28] But before they could begin, the British government first needed to do one more test, once again at sea.

At home, Churchill had bowed to public pressure, and ordered the Medical Research Council to study the genetic effects of radiation in man. Worldwide protests were growing, and Communist infiltration was feared and hunted in every public sphere. Penney's team had the simultaneous problems of creating a device

The ship's company of destroyer HMS *Diana*, pictured in 1956 before she was ordered to twice sail through the fallout of Operation Mosaic. (Courtesy of wireless operator Bob Malcolmson)

so big it could flatten a city, and the need to do it before an imminent ban on the experiments needed to make the weapon work. Britain had to do more atomic tests as a matter of urgency. And the next was to be the biggest, and most technologically advanced, bomb they had ever exploded.

Three years after the black mist of Totem had settled, the destroyer HMS *Diana* and her crew of almost 300 men were ordered to sail from Portsmouth on a secret mission. The aim was to see if they and their ship could function in the event of a nuclear explosion.[29] Churchill had been replaced by Anthony Eden, who agreed the threat of apocalyptic retaliation was all that would keep the country safe. *Diana*'s mission was an intentional human experiment, although British governments have never used that phrase. It was codenamed Operation Mosaic.

Diana's captain John Gower was a hero of the Arctic Convoys, took part in the attack on the Nazi battleship *Tirpitz*, rescued 80 men at Dunkirk, and was awarded the Distinguished Service Cross for his part in the D-Day landings. His nephew, David, grew up to be captain of the England cricket team in the 1980s. After he retired, this war hero began campaigning for the crew of HMS *Diana*, and wrote a seven-page account of what befell them on his watch.[30]

In early 1956 he joined the ship at Devonport docks in Plymouth, where she was fitted with 'a large metal fin designed to be equipped with scientific instruments and towed in a vertical position from the forecastle. The instruments were to be inserted to measure the level of radioactivity in sea water'. They spent March in sea trials, learning how to tow the fin, but Gower said it kept getting stuck under the ship. Nevertheless, *Diana* sailed to Australia, via Gibraltar, Malta and the Suez Canal. After entering warmer waters, Gower told the sailors their destination was Monte Bello, where there would be two nuclear tests.

Gower took a dim view of the boffins who travelled with them, saying they 'lacked shipboard experience'. And that pesky fin was junked. He wrote: 'The fin proved to be useless and was replaced by polythene sheeting and weights, designed by the Chief Scientist on board and the Gunnery Officer. This proved to be entirely satisfactory and the original apparatus was landed in Singapore as scrap metal.'

But while he alone knew the ship's task was to gather information on fallout, Gower was left in the dark about the reasons for it. He said: 'It was only years later that I read the illuminating power politics which formed the backdrop to our adventures... The purpose of all our tests at that time was to show Washington (and Moscow) that we were a full member of the nuclear club. Time was running out before a worldwide ban was imposed and Britain wanted the hydrogen bomb. Perfection of an operational trigger device was of great urgency.' Mosaic was to be a vital halfway house, that could in future be used to turn an A-bomb into an H-bomb.

Gower said the trial was rushed: 'Mosaic I and II were organised with great urgency... How much radioactivity could a ship withstand and remain operational? HMS *Diana* and her crew was made available to provide the answers.'

His crew and ship, he said, 'were ordered to steam deliberately through two atomic fallouts. The MoD's attitude at this time was that radiation from these tests would have no measurable effect on those taking part. This statement was made in

1953 and was still maintained in 1983.' What the crew never understood was why something expected in advance to have 'no measurable effect' required a test to prove it.

Gower was ordered to Monte Bello for May and June, when prevailing winds would blow from the west. By July, storms would render the trials impossible. As *Diana* travelled south, the crew practiced pre-wetting the ship with seawater to prevent fallout settling, and rehearsed being locked inside two 'citadels' below decks, one holding 207 men and the other 95. In order to keep radiation out, these areas were pressurised with air pumped in from outside. But tropical conditions meant the lockdown was not as easy as it sounded, and Gower said that 'some suffered fatigue, faintness, headache, respiratory distress, abdominal discomfort and claustrophobia'.

He added: 'The engineers had a gruelling time in the heat of the Red Sea and Indian Ocean. They were required to wear space suits with air intakes. When connected to a filtered air supply, and fully blown up, they looked like Michelin men. The legs of their suits quickly filled up with sweat. The rest of the ship's company wore working rig, seaboots, and carried gas masks.' Scientists gave regular 'reassuring lectures' about nuclear weapons, and the crew were shown unreleased footage of Hiroshima and Nagasaki.

After sailing 10,000 miles in five weeks, *Diana* arrived and anchored three miles off Monte Bello. A fleet, led by HMS *Narvik*, was already there, and scientists came aboard to assess *Diana*'s readiness. Gower wrote: 'Their report, which I did not see at the time, read "general impression was everybody understood the basic problems and how to tackle them but they have fallen short in the apparatus and tools to do the job". This was followed by the usual disclaimer: "We cannot, however, assume any direct responsibility for the radiological safety of *Diana* since this is outside our terms of reference." So doubt did exist.'

The plan was for Mosaic I, a 15 kiloton device, to be detonated at the top of a 100ft tower built on Trimouille Island, just a few kilometres from where *Plym* had been vaporised four years earlier. *Diana* would chase its cloud, and sail amid fallout. She would stream the redesigned fin behind her, and lower a small whaler with instruments, to be left bobbing on the waves to take measurements. Further devices would trail from *Diana*'s stern to see how far under the water radiation could be detected.

Meanwhile, the ship's engineers would check radioactivity in machinery spaces, engine room, generator, and pumps for drinking water. While most men were to be kept below, some would be above deck acting as watch keepers in 35 minute shifts, 'which is all they could endure in their space suits', said Gower.

Mosaic I was exploded at 11.51am local time on May 16. Gower wrote: 'The ship went to action stations at 1220 and to shelter stations at 1305. The expected arrival of 95% fallout was 1325. First readings, however, showed us as being only on the fringe of the fallout. No real effect was experienced until 1915 and fallout ceased at 2100. The monitoring staff, fully protected, left the citadel at 2135.' This small team took radiation measurements, which were around 5 miliRoentgens - about the same

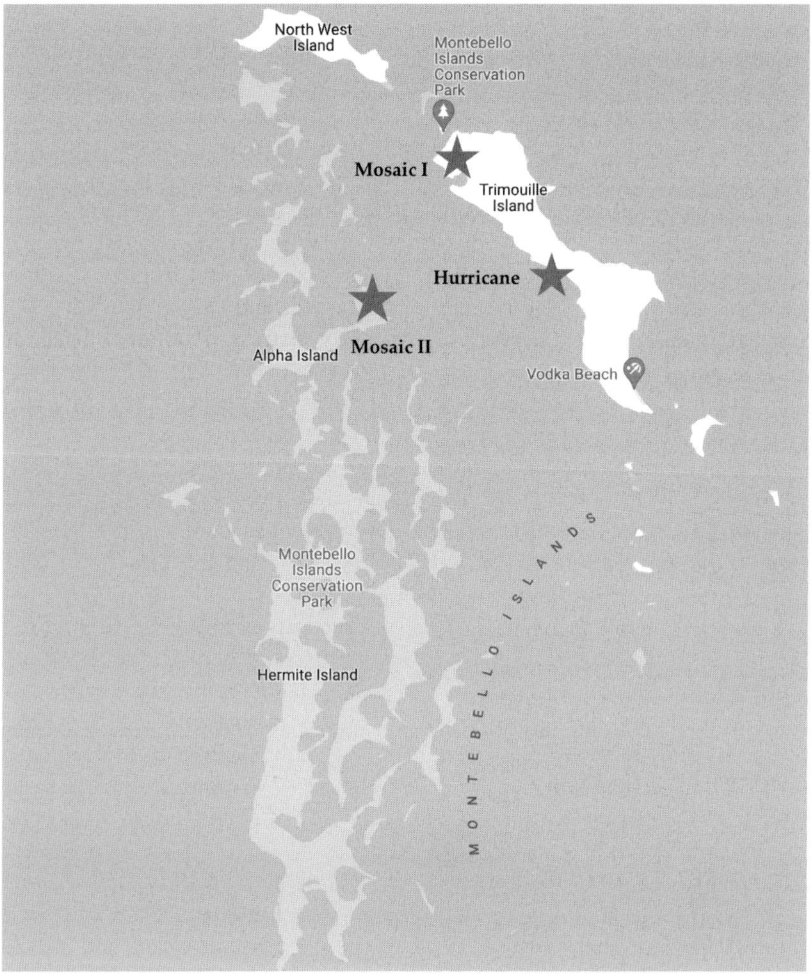

A map of the Monte Bello islands, showing the three locations of Operations Hurricane, Mosaic I, and Mosaic II. (©Google 2024)

as two chest x-rays. Sailors were ordered to scrub down the decks an hour later, and they were not declared 'clean' until 1am.

After 13 hours of cramped conditions, heat, sweat and stress, and no ability to smoke, the men were keen to leave their confinement. But outside, there were other issues. 'When recovering the scientific data from the whaler it was discovered that the boat was "hot" so we sank her by ramming. We found also that seaweed and debris that had accumulated in the circulating filters was also "hot",' Gower wrote.

He had hoped his crew could get a few days' shore leave in Fremantle, but the Australian authorities refused entry for fear of contamination. They sailed to Singapore instead, via a small atoll in the South Pacific named Christmas Island where they left stores for a handful of troops, for reasons that would become clear later. They arrived back at Monte Bello for a second test on June 7, but the mission was delayed by bad weather. While they waited, the crew grew idle. Gower wrote: 'After several false starts the ship's company were beginning to get bored… to counteract this we staged every conceivable dog-watch pastime to while away the time.' While sailors twiddled their thumbs, the scientists were fretting. Mosaic II was the most powerful device they had yet designed. It had to take place, and soon, whether the wind was right or not. There was political pressure from home: the MRC report was published on June 19, and had decided that radiation was very bad for humans indeed. The *British Medical Journal* said its main conclusion was that 'as all such radiations are potentially dangerous, their use should be the subject of constant and close scrutiny, and that adequate justification should be required for the employment of any source of ionising radiation on however small a scale'.[31] It said health effects observed after the Japanese bombs had included high rates of leukaemia, miscarriages, and birth defects.

Gower recalled: 'The pressure was now on the scientists. In order to obtain the trigger they so urgently needed for the H-bomb, it was vital that Mosaic II should take place. To cancel or postpone the test would involve a great waste of time and money and the decision to proceed was finally taken despite the risk of adverse

An official image of the 98-kiloton Mosaic II blast, in the seconds after detonation, on June 19, 1956. (Wikimedia Commons)

weather conditions.' Not only was the weather unfavourable, but this second blast would be huge. 'Unknown to us at this time, Mosaic II was to involve a considerably larger device. With a yield of 60 kilotons it was to be the most powerful A-bomb used in any of the [UK's] 21 tests,' said Gower. 'As it later emerged, however, even this information was inaccurate… it was not until 1985, almost 30 years after the event, that the MoD eventually admitted that the actual yield of the bomb was 98 kilotons, seven times more powerful than the bomb used at Hiroshima.'

The device, again on top of a tower and this time on Alpha Island, was exploded at 10.14am on June 19. *Diana* was 97 miles away, and Gower watched the flash from the ship. He said: 'The fireball was twice the size of Mosaic I and climbed a lot faster. The explosion when it came was a colossal double crack heard 200 miles away in Australia.' *Diana*, waiting out at sea to chase the cloud, was expecting fallout to arrive around three hours later. Gower wrote: 'Both citadels were pressurised by 1245 and the first traces of fallout were detected at 1325. By 1440 we had recorded radiation in excess of 10 miliRoentgens but this level had dropped to 0.4 by 1845.' Decontamination parties left the citadels and washing down the ship was finished shortly after midnight. They found one pre-wetting hose, supposed to be pumping water over the ship's surfaces throughout the lockdown, had broken, and the deck was heavily contaminated. Gower said the sheer size of the explosion meant the ship had other problems. 'Other areas were found to have an unacceptably high radiation count,' said Gower. 'Repeated efforts to reduce this count were unsuccessful so these areas were roped off for 20 hours. Finally we recovered the whaler and sank it by rifle fire, thus completing our task.'

For both tests, the crew's only safety equipment was overshoes and gauze masks. Only those keeping watch had been given the 'space suits'. And the capricious weather had not helped. 'Later we were to learn that during Mosaic II the radioactive cloud began moving away from the Australian mainland but then altered course and drifted back,' said Gower. 'This prompted an angry signal from the Prime Minister of Australia to the British […]: "What the bloody hell is going on? The cloud is drifting over the mainland".'[32]

The *Adelaide Advertiser* reported that the cloud was spotted in Perth, but Australian Minister for Supply Howard Beale said there was 'no cause for alarm'. The paper told its readers: 'Mr Beale said some cloud had drifted inland temporarily, but this need cause no anxiety.'

Since the Second World War, Australia had provided the raw material for Britain's atomic programme, and the Outback was peppered not just with small uranium mines, but inquisitive miners who were radiation experts and armed with their own Geiger counters. After Mosaic II, the *Advertiser* reported the claims of one who had found a 'normal count of 15' in the morning and 'the counter leaped to 2,000' after subsequent rainfall. Reporters trying to cover the story from Woomera found the government had cut off the local telephone exchange. In an effort to keep a lid on it, Titterton's safety committee issued a statement that 'we cannot over-emphasise the fact that the whole operation was carried out with no risk to life or property on the mainland or elsewhere'.

Archie Hart, pictured aged around 18 before he was ordered above decks while *Diana* steamed through fallout. (Courtesy of Archie Hart)

Archie Hart was an 18-year-old stoker who had joined *Diana* at the start of the year. The son of a Warrington factory worker, he had been a junior railway clerk before being called up for National Service. He says the crew were told they were to sail through the fallout only a few days beforehand, and received no specific safety warnings. Their training was about how radiation was measured in Roentgens, rather than its impact on living things.

'When the bombs were detonated I was up on the deck of the ship. I remember that the exact place where I was standing was enclosed around the sides, so I could only see the sea and the sky, not the site of the detonation,' he said. 'When both bombs exploded the first thing I saw was a very bright light on the sky, like lightning. This was followed by a loud bang.' Archie remembers clearly he had no film badge, no advice about contamination, and while he had to wash down other members of the crew, he didn't get a shower himself. While others wore 'space suits', he wore only a cotton hood over his head and shoulders.

Three years afterwards, Archie developed a large, fatty lump on the back of one hand. 'If you knocked it, it would go up my arm, and if knocked again it would go back along my arm to the back of my hand,' he said. 'There's 100 all over my body now, some the size of tennis balls. I can't do the dance of the seven veils anymore, because my body's an unsightly mess.' He suffered boils while on *Diana*, and in the years since they've recurred, along with cysts, unexplained rashes, and suspected tumours. Aged just 65, he had 18 inches of his bowel removed after developing cancer. His wife Noreen suffered miscarriages, but they went on to have five healthy children, although his eldest daughter was unable to have more than one child. Nevertheless, with every grandchild and pregnancy he worries about a radiogenic deformity.

'No-one sought our permission to use us in an experiment and we did not volunteer,' Archie said. 'We were guinea pigs, sent into danger by men who were sat safe behind a desk in Whitehall. Even the officers on board didn't know much more than us. What they did to us was morally wrong, and what is morally wrong cannot be politically right. Their cavalier attitude in the years since is causing problems to this day for the generations that follow.'

Other vessels from the fleet stayed close to the blasts, and crews were ordered into Ground Zero, wearing only shorts and boots, to collect military equipment that had been left onshore, like generators and jeeps. Veterans later recalled seeing steam rising from the sand, and said walking on it was like 'a skating rink' because the surface had been turned to glass. Dead fish floated on the water.

Before Mosaic, Commodore Hugh Martell aboard the *Narvik* cabled the rest of the fleet: 'There is a possibility that, in certain cases, there might be some slight radiological hazard arising from eating fish that are caught in the immediate vicinity of the Monte Bello islands... This precaution is not intended to spread alarm and despondency, but, in an operation of this nature, it is essential that full precautions should be taken from the start and relaxed in varying degrees later as more information becomes available.'[33] He said they were seeking confirmation that seafood could be eaten, but until then, no-one should go fishing. The message did not get through to every sailor.

Roger Grace was an 18-year-old National Serviceman aboard *Narvik*. He said: 'Not only did our government lie to the Australians about the bombs we were going to explode, but they left us to live in the fallout for four months protected only by cotton shorts.' For the first test, with the rest of the crew, he was 'mustered on deck in shorts and shirts, and told we were about six miles away. [The second blast] was different… this time we were 12 miles away – they knew it was going to be more dangerous. Instructions were given for us to stand with our backs to the detonation and put our hands over closed eyes. The flash was so bright that I could see the bones in my hands. Like an X-ray, it was as if a searchlight was being shone at me. The heat was felt on the back of my neck as if an electric fire was being held there, and the blast whipped my clothing away from my body like a flag in a gale. It was big.'

A few hours later he was told to take a few scientists in a small boat and land them on a beach close to where the towers had been. He said: 'I was just in shorts. They were in shirts and trousers.' Senior officers warned him some places were radioactive, but didn't say which bits, and there were no warning signs. Roger said: 'I was given just four film badges to measure radiation, but they were attached by a safety pin so of course I lost them all in the course of a day's work.'

The men were not allowed to take a camera for security reasons, but Roger persuaded the ship's photographer to let him have a few of the official copies as a memento, and hid them in his kitbag. They showed him and his friends lounging in the beer tent and swimming in a lagoon that modern health standards would never have let him near. Their makeshift pub was just four miles from Ground Zero. Roger said: '[It was] built for us by the Royal Engineers to keep the men happy. They also built a desalination plant. We'd go for a walk, drink beer on the beach and then have a swim. The shark net on the beach was the only safety precaution we had. We ate fish we caught in the lagoon, we didn't have a care in the world. But they knew.'

On his return home Roger left the navy and joined the police force, but in the 1970s began to develop cancerous lumps on his face called 'rodent ulcers' which had to be removed. He has had half a dozen cancerous skin tumours cut off his back and legs, suffered bowel cancer, and developed angina and heart problems. All of the illnesses can be caused by radiation – but he was repeatedly refused a war pension, on the grounds that the MoD didn't 'recognise' any link to his service. 'I've been lucky,' he said. 'Unlike so many others, my wife didn't have miscarriages and my children and grandchildren were mercifully born without the birth defects others have in their families. But those islands were only a few miles wide. We were never far from the test site… they will not hold their hands up, but they should. They simply wouldn't be allowed to do the same thing today. I firmly believe the government treated us as guinea pigs.'

Diana returned to Singapore, where a radioactive boiler was repaired and the ship repainted from bow to stern. But the crew's achievements had blown away with the fallout; the ship didn't even get an Admiralty telegram saying 'well done'. As the unheralded crew sailed home, they were again refused permission to dock in Perth. They stopped instead at Malaysia and India, then were stuck for three months in Aden, now part of Yemen, as the Suez Crisis erupted. *Diana* helped to sink an Egyptian frigate, rescued survivors, and for that was finally sent signals of congratulation from the First Sea Lord. They returned to Plymouth in early 1957, having sailed 25,000 miles in just a year. Her crew had spent most of it on board a contaminated vessel, with close to zero medical advice, check-ups, or protection.

They returned to near total disinterest. The bomb tests had been all but forgotten amid a Middle East crisis, rising oil prices, and Eden's resignation and replacement by Harold Macmillan. But something about it still niggled at Gower. Why, having expressed a wish to know what happened to servicemen exposed to fallout, did top brass never follow up? He wrote: 'What puzzles many of us is why, on conclusion of the trials, no effort was made to see what effect the nuclear explosions had both physiologically and psychologically on the men who witnessed them and, perhaps more important, on those who went through the fallout. Yet we had no follow-up medical checks of any kind. From February to August 1957 *Diana* remained in home waters. This does seem an appalling waste of a unique opportunity. Mosaic II, I believe, cost in the region of £1.5m pounds [£364.3m in today's money]. Why not get value for money by testing everything - men as well as weapons?'

Perhaps the answer lies in what has been uncovered since. In a declassified 1955 memo before the tests, Penney wrote to the head of the Atomic Energy Authority, Edwin Plowden, saying the first blast should be between 10 to 20 kilotons, but they had no idea how big the second one would be. He said: 'We do not know exactly what the yield is going to be because the assembly is very different from anything that we have tried before. We expect that the yield will be 40 or 50kt but it might just go up to 80, which is the safe upper limit.' At the time, the British government said both of the bombs were less than 20 kilotons, the size associated with the weapons dropped on Japan. But after the stories of Yami Lester and others caused public outrage, the Australian government ordered a fresh analysis. In 1983 it reported the true size of the second Mosaic test was 60 kilotons, making it four times as powerful as the bomb which destroyed Hiroshima.

We now also know the huge yield was achieved with a core of uranium-238, which has a decay half-life of 4.5 billion years. It would take more time than the Earth has existed for the weapon's ingredients to be just half as toxic to life. It has always been the government position that all its nuclear tests were designed to consume such materials in the blast, and for the explosions to happen in a way that minimised fallout. But in 1984, the *New Scientist* reported: 'Mosaic II was mounted on a tower 30 metres high. For a 60 kiloton weapon, the tower would have to have been a kilometre high to avoid radioactive debris being flung into the air as the fireball touched the ground.'

Put simply, Mosaic II was a massive dirty bomb. One AWRE study thought it could have been as big as 98 kilotons, which is more likely for a weapon that had a boosted yield, as this one did. In either case, it was detonated too close to the ground, and is highly likely to have created unexpected fallout. Gower wrote that the British still insisted 'no-one was exposed to any significant health hazard, therefore no tests were necessary' on the crew. The logic amazed him. 'Without testing,' he asked, 'how is it possible to be sure there were no hazards?'

Gower and his crew mates met regularly for reunions and began to notice disease and deaths which came before their time, and former comrades with sad stories about deformities in their children, and later grandchildren. He wrote: 'Even if one were to accept that the statistics available… have not, so far, proved anything, it must be pointed out that our experience in HMS *Diana* was very different from that of later tests. We were required to steam through the fallout from two nuclear explosions, to deliberately contaminate our ship and to continue to serve in a ship, parts of which had been unacceptably radioactive. The MoD's attitude remains as always "if you think you have a case, prove it!" No individual can do this. One has to accept that it is a battle lost.'

Gower died in 2007, aged 95, physically unharmed by his experience. But the rest of his crew was not so lucky. In 2008 *Diana*'s wireless operator Bob Malcolmson joined a class action lawsuit against the MoD. 'The explosion was tremendous… We turned our backs, covered our eyes with our hands. I had my eyes open and I could see the bones in my hands, even with my back to this thing,'

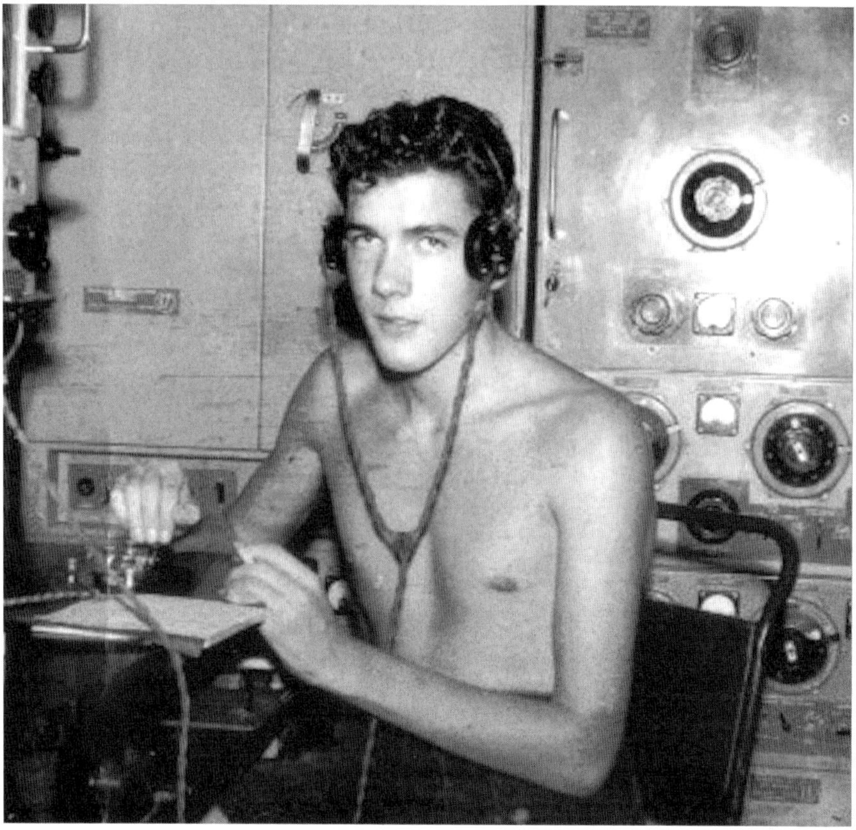

Bob Malcolmson, aged 18, at his station in HMS *Diana*'s wireless room before taking part in Operation Mosaic. (Courtesy of Bob Malcolmson)

he said. 'Several chaps lost teeth, and others lost their hair. So a lot of wives and sweethearts waited in Devonport to welcome back bald fiancés and boyfriends with a few teeth missing.'

Bob, originally from Belfast, joined the navy as a boy sailor at 15. He was 18 at Monte Bello, and a junior officer when he left the service in 1970. Four years later he was diagnosed with polycythemia rubra vera, a type of cancer in which the bone marrow produces too many red blood cells and which is known to be caused by radiation. In 1990, after a long argument, the MoD admitted this condition was due to service and made a one-off payout of £2,470 - worth £5,656 in today's money.

In the decades that followed, Bob went through 318 different procedures to remove excess blood. He had a heart bypass, glaucoma and cataracts in both eyes, a deep vein thrombosis, developed the incurable circulatory condition Raynaud's disease, had four operations to remove skin cancers, eight operations on his leg including an arterial bypass and removal of an aneurysm, two stents, and had part of his penis cut out to remove a malignant tumour.

In 2006 Bob finally won a full officer's war pension, after the MoD agreed all his conditions were linked to service, and moved to Alicante in Spain. There, he was diagnosed with terminal lung cancer, the risk of which can be increased by radiation. In his final interview, Bob said: 'I have a daughter, granddaughter and great-grandson who are all healthy, so at least I do not seem to have passed it on in my genes for which I am grateful. I loved the navy, had a brilliant life and am still enjoying it. But I challenge the government to look me in the eye and tell me no-one was injured by Britain's nuclear tests. I have seen too many of my shipmates die... We are asked to prove we were irradiated, which is virtually impossible.'

Dozens of scientists were sent into danger as well. One of them, Dr Stuart Ellis, stood on *Diana*'s deck, taking samples, while the crew were locked below. The year afterwards he lost all his hair, and was later diagnosed with leukaemia. It was described by his doctor as 'the most unusual presentation of multiple myeloma I have ever seen'. He died in 1989, just a day after his diagnosis. His widow

Dr Stuart Ellis was a civilian scientist ordered to take radiation readings on the deck of HMS *Diana* as she sailed through fallout. He later died from 'the most unusual presentation of multiple myeloma' his doctor had ever seen. (Courtesy of the Ellis family)

Edna said: 'He started losing his hair straight away and in a year it was completely gone. For years he was always breathless and couldn't take any exercise. He kept going to the doctor because he felt so bad. One day they sent him home and two hours later rang me and told me to bring him straight back. They found tumours in all his organs. I stayed with him that night and he died the next day. He was only 63. Our daughter, who was born 2 years after the tests, had a curved spine that had to be straightened with metal plates, and she had a daughter with a hole in her heart. Her son was very premature and spent his first few weeks in intensive care.'

A report written at the time of the tests by Dr Ellis, known to friends as Sam, says the ship was exposed to alpha fallout. These particles of dust or debris are sucked into a blast, and fall back to Earth carrying a faint radioactive charge. It can be pernicious, able to lodge in the human body for decades, and is carcinogenic. He wrote: 'The risk is omnipresent. Thus dirty fingers may convey the contamination to the lips or food.' He added the average sailor could find the 'ship's fittings and structures, tools, ropework etc could be dangerous to him through his fingers and mouth for days or weeks after the ship was apparently "clean".'

Dr Ellis left the ship after the first explosion, ordered to come straight home with his samples so they could be examined. Declassified documents reveal he carried one hundredweight - or about eight stones - of radioactive material in a container onto a commercial airliner. 'I collected him from the airport with his bag and a briefcase. We were living in a flat in Fulham and his case stayed in the hall as it always did until he went to work on Monday. He said it had his samples in it,' Edna remembered. 'It wasn't until Stuart died I realised there must be a link with Monte Bello. If the men were sent below decks they must have known it was unsafe - and he wrote a log for the navy of everything that happened. But it's been 50 years and they still deny it all.'

Dr Ellis, a civilian based at the National Physical Laboratory, was the only scientist whose family were known to have later taken legal action against the MoD. Edna said it was because her daughter was so ill, and her husband had suffered so greatly. 'In the last year of his life he faded away to almost nothing. He went to the doctor and they told him he was fine,' she said. 'Two weeks later he was dead.'

For the scientists, Mosaic was a success. Atomic weapons use fission, or splitting atoms, to release energy. In simple terms, you take a highly unstable metal - such as plutonium - and surround it with explosives, then wrap this core in a hard outer jacket that can withstand huge pressure. When the explosives are detonated, the blast bounces back inwards, splitting already-volatile atoms apart to release huge amounts of energy. With an H-bomb, the science goes a step further - and uses one atomic bomb to detonate another. Imagine you rebuild your original device, with explosives, a radioactive core, and a tough jacket. Around this you then pack yet more material, with another core, and a second jacket. The first bomb tears apart the atoms. This sets off a second reaction, and this time, it smashes the atoms back together again in a process called fusion, the same power that creates the Sun.

For success, the detonations must be precisely timed to within a hundredth of a second, and the process, power, and yield can all be affected by what materials are used, and how much. The Mosaic blasts were planned as "boosted fission" weapons, in which a small amount of fusion was used to boost the yield. Mosaic I had failed, but Mosaic II had worked better than expected. Its power was increased by being formed from two explosions, the double-crack Gower heard at the time.

In 1955, Eden had misled Menzies, telling him Mosaic would involve 'atomic explosions with light elements as a boost'. He said the blast would be twice that of Hurricane, but this was useless information as no-one outside the UK government knew the yields. But they were also intending to contaminate *Diana* and put her crew in danger. As Martell later admitted: 'My job was to see that the ship sailed at the right time and that she went into the right area to receive maximum fallout.'[34] Planning documents from 1956 state that 'this was the first operation by one of HM ships to gain first-hand experience about fallout and contamination' and it was a chance to see 'what hazards if any to engineering personnel are to be expected as a result of passage through fallout from atomic weapons'.[35] One memo shows the Director of Physical Rcscarch suggesting 'low' exposure standards for the crew that were equivalent to 13 years' worth of normal background radiation, higher doses equal to 87 years' worth, and a special dose the same as 219 years' worth. Medical Director General of the armed forces, John Morley Holford, agreed to the rules, saying he 'appreciates that in these infrequent, and very expensive, operations it is reasonable that some slight risks should be incurred by personnel rather than that important records and observations should be lost'.

Papers written by the scientists after the blasts admitted deciding to sail *Diana* through the fallout with her vents open 'so that we could find out precisely what contamination was drawn in, but we could not tell how serious a problem this might produce'. They said 'levels of loose contamination were well above the health tolerance throughout both the machinery spaces in use' and 'there might be some danger from the ingested hazard at some later stage'. They found fallout in the vents, on boots, ladders, and decks. A meter used to detect radiation 'appears much too insensitive' and worked only for contamination several hundred times greater than the maximum that had been set. One report states: 'The principal conclusions are... that the resulting contamination is mainly loose, that it is easily transferred to the lungs and stomach and must therefore be considered a hazard, even though the external dose rate is very low.'

Commodore Martell wrote a report falsely stating the cloud never crossed the Australian mainland, and was later appointed CBE. He retired a vice admiral.

In 1999, an MoD civil servant wrote to one veteran saying 'the confusion over the track of the cloud' was due to communication difficulties with Australian officials. He added: 'The participation of HMS *Diana* in Operation Mosaic was meticulously planned over a period of nearly a year... due regard was shown to the exposure of her crew to ionising radiation and dose limits were laid down.' But in 2006, scientists confirmed that the exposures described as low at Mosaic were high enough to cause cancers.

HMS *Diana* was decommissioned in 1969, and sold to the Peruvian navy where she served until 1993. She is the only ship to have survived two nuclear bombs, although the same cannot be said of her crew. Today fewer than a dozen of them are thought to still be alive. More than half have died from tumours, and they show an increased risk of dying from prostate, bladder, kidney and blood cancers. So far as anyone knows, and as Captain Gower always feared, no medical checks were ever conducted on those who took part in what was Britain's first human radiation experiment. It would not be the last.

Britain had the bomb, but it still wasn't good enough. The only way to deliver it was using gravity to physically drop it on a target. What top brass wanted was the capacity to fire it from home turf, making it defensive rather than offensive, and reducing the need for risky bombing raids that might see the plane crash in the wrong spot. Penney began to feel he was leading the fight for the entire Western world's way of life. In a talk at Melbourne University in August 1956, he told government delegates that Russia's atomic success was 'frightening' and Western universities should demand and get more resources to compete with the state-sponsored physicists of the Eastern bloc.

Sounding more like a politician than a mathematician, he said: 'I believe this is true, important, and urgent. We are living in difficult times. Russia has challenged us in the military sphere, and is now entering most successfully the economic sphere. We are inculcating into many countries our system against theirs. We must help these countries in technological and cultural ideas… money spent on universities is not merely an investment, but a crying requirement if we are to meet the challenge of our day.'[36]

Focus moved back to Maralinga, where troops built a military town in the desert, consisting of barracks, offices, mess halls, laboratories, blast-proof bunkers, roads, and water-purifying facilities for a new series of tests codenamed Operation Buffalo. The new settlement was 14 miles from the blast zone. Aboriginal people were allowed into the camp to sell boomerangs, but most were forcibly relocated. Others were chased out of the 100km exclusion area by patrols, but neither side were able to communicate with each other, and may have been equally ignorant of the risks.

When news of the plans broke, there was little media opposition. Trade unions organised small public protests, and Penney took on a propaganda role. He cabled London: 'Greatest problem is to relieve public apprehension about radioactive fallout. Definite progress has been made but there are small hostile groups trying to make mischief. I am confident that we can handle these matters.' Meanwhile, the abandoned Emu Field - where tents, food and tools had been left where they lay since the Totem fallout drifted towards it - was visited by troops seeking souvenirs and equipment, and indigenous locals, none of whom were warned about the dangers of bringing away a set of spanners or can of peaches.

The British were increasingly cagey with their Australian hosts. In December 1955 Penney wrote a memo to the MoD's chief scientific adviser, suggesting evidence from cloud sampling missions at Buffalo be withheld. He wrote: 'We think it is likely that the Australians will ask us for filters [used to record radioactive particles within the cloud]… While I am not keen on giving them samples, I do not see how we can refuse. They could, of course, fly planes of their own or they could most easily take contaminated soil particles from the close-in area.' He said these particles were very different from those in the cloud, and the Australians might not realise that they were more crucial than 'a few shovelfuls of dirt from the crater'.

He concluded: 'I am recommending that, if they ask us, we give them a little piece of the filters, but we wait a few days so that some of the short-lived key isotopes have decayed a good deal.'[37]

Buffalo would involve four atomic tests, codenamed One Tree, Marcoo, Kite, and Breakaway. The aim was to make warheads lighter and smaller. Two were on towers, and for the first time one was dropped from the air because, in the words of one official, 'it also provided the RAF with the opportunity to drop a nuclear bomb'. Another would be exploded at ground level to maximise fallout, in the hope of producing data on 'cratering and other weapon effects' that could be shared with the Americans, to increase co-operation.

After Hurricane, the device that had sunk the *Plym* had been turned into a warhead called Blue Danube. But it was big and heavy, and in 1953 the armed forces asked for a smaller one. The redesign was called Red Beard, and now both were to be fired for the first time. In press conferences, Penney assured the Australian media that 'there would be no danger of any kind involved in the explosions'. What he did not say was that men were to be part of the experiment.

The military had decided it was important troops see first-hand what happened at and after a nuclear explosion. Buffalo would be witnessed by servicemen from the UK and Commonwealth, offered double pay as enticement. At a War Office meeting in Whitehall in August 1955, it was felt that 'the most should be made of this opportunity' and 'that any extra use made of the trials must be regarded as a bonus'.[38] In a sign that they knew these men would be exposed to fallout, the meeting discussed decontamination procedures and that it would be necessary for these troops to be kept separate from others for fear of 'cross-contamination'.

The idea was not universally popular. A memo written on November 29 by Group Captain Stewart Menaul of the Atomic Weapons Trials operations team said: 'The Army has suggested that they be allowed to send 250 selected soldiers to the Buffalo trials… There has not been wholehearted support for this venture either by the scientists or by members of the other services.' There were logistical problems with a 12,000-mile supply line, and extra men would stop the scientists doing the work that needed to be done. He concluded soldiers would have to be familiarised with the weapons, but said: 'I do not consider however that Operation Buffalo is an appropriate experiment at which service indoctrination could take place.'[39]

His objections were ignored, and a series of experiments were planned. In one, 120 man-sized dummies with measuring devices in their heads and stomachs were strapped into trucks and tanks, or arranged as though sitting, lying or kneeling on the ground, at varying distances from the bomb. They were dressed in uniforms of the three services, and between them and the explosion were placed a series of glass windows. Many victims of the Japanese attacks had been shredded by glass shards, and the British wanted to find out what damage the Buffalo tests could produce in a city. The live troops in the Indoctrinee Force were to be taken on a tour of these dummies after the blasts, to see for themselves the likely effect on flesh.

They would fill in before-and-after questionnaires to assess changes to knowledge, opinions and morale. But it wasn't enough. An Army Operational Research Group report to the Chiefs of Staff said: 'The object of their indoctrination was to give them a better idea of the nature and possibilities of nuclear weapons and effects than could be inculcated by lectures and "cold" or faked demonstrations alone. Also they could pass on their experience to other members of the Armed Forces at the conclusion of the trial.' It said 'first-hand experience of a nuclear explosion must have a powerful effect on one's attitude to this type of warfare' and it would be a good idea to assess the impact on troop morale. But it stressed the need for a 'realistic exercise' because 'the indoctrination of a force consisting entirely of officers of all ranks, up to and including Major-General, who merely stood up and watched an explosion nearly six miles away, did not quite lend itself to the solution of this problem'.[40]

An official statement was issued, saying that '250 officers from the armed forces of the United Kingdom, Australia and New Zealand will be exposed, at a safe distance, to the flash heat and blast effects of an atomic explosion. A conducted tour of the firing area will be made before and after the event to enable indoctrinees to observe and appreciate the effects of explosion on the ground and on items of Service equipment, vehicles and structures.' It said the men would be four and a half miles from the blast, and would 'pass through health controls and don protective clothing to enter the contaminated area'.

In August, Australian newspapers reported that although it would 'be the first time Britain has exploded an atomic weapon on the ground' it was 'aimed at discovering its effects on military tactics and civil defence structures'. Penney told reporters the indoctrinees 'will have the opportunity of hearing, seeing and generally experiencing an atomic explosion from a safe distance'. He boasted: 'I am certain there is no danger of radiation blast or fallout from the tests.'

A few months later, a letter from the War Office to AWRE warned that the tower blasts would not limit fallout at all, saying: 'Contamination from firings on lower towers will be more severe.' Eden, before he had left office, also raised concerns. A 1955 memo from the Lord President's office says he wanted to 'look into the implications of limiting nuclear weapons tests to one a year... the purpose of limiting tests would be to limit the genetic hazards'. It suggested the nuclear powers impose a cap on megatons, as 'then the genetic hazards would be directly limited'. But it went on to discuss finding a way around any limit, and ends with a handwritten note: 'This seems a thoroughly sensible suggestion. I will put it to the Prime Minister.'

In 1956, a press release was drafted to calm the Australian public, and mentioned 'biological investigations' that would take place in the wake of the blast. It was discovered and reported by the *Sunday Mirror* in 2011, and appears to have been declassified by mistake, as the front of the file states it should not be opened until 2032.

Marked Top Secret, it reads: 'The United Kingdom and Australian Governments have agreed that investigation of biological effects of atomic weapons shall be carried out during the trials at Maralinga in 1956. The possible effects of the ingestion of radioactive fallout (by men and animals) will be among the subjects studied.'[41] The draft had the words 'by men and animals' crossed out in pencil. The version that was finally released mentioned only that 'a few sheep and small animals' would be studied, and that 'elaborate arrangements for the welfare of the animals are being made'. There was no mention of human hazards. A separate press release from the Foreign Office, emphasising to the Australian public that the purpose of these explosions was finding out how to protect them from enemy attack, came with a note telling the Australians to put it out when 'it will be lost in the weekend flood of sporting news'.

An internal MoD telex from July 1956 urged those on the ground to get the Australian safety committee to tell a flat lie to their citizens about the the MRC report a year earlier, which found radiation was linked to an increase in disease and birth defects. It says: 'We do not want you to release any statement on genetic effect or on radioactivity or strontium pending the arrival of Penney. If you have to, a safer interpretation of the MRC report in the last sentence of paragraph 4 would be "has not shown an increase" rather than "shows an increase".' This time, the cover-up had begun before the weapons were even detonated. And those due to be indoctrinated knew nothing of it.

In the event, there were 283 indoctrinees - 172 British officers, 100 Australian officers and lower ranks, 5 officers from New Zealand and six civilians. They messed and bunked apart from others at Maralinga, and wore full uniform at all times, despite the heat. It gave rise to gossip among the rest of the troops, who wore little more than shorts most days and claimed to have heard eerie wails at night, that these were 'mental defectives' brought to have experiments conducted on them. The rumour circulated long enough to be reported by *The Times* in 1984, but was eventually found to be baseless.

The men were given six days of lectures, and tours of the testing grounds with scientists. The talks, some given by Penney himself, included 'the effects on personnel of heat flash, blast effects, ionising radiation, biological effects of fallout, radioactive contamination, health physics and health control and radioactive decontamination'. There were reports more than 120 Aboriginal people, barefoot, bareheaded, and in various states of nakedness, had been seen on the range six days before the first test was due. A native patrol officer was sent to check, but without a vehicle of his own, or a radio. Official census maps drawn up at the time marked the area 'vacant'.

A delay in firing the first weapon was revealed at the end of August 1956 by reporter Chapman Pincher, who said the team had hit 'unexpected difficulties' that meant a fortnight's delay. Another report claimed that 'with 1,500 men waiting at Maralinga for the explosion it is obvious each day's delay is costing at least £3,000 in pay alone' - or £77,000 in today's money. One Tree was pencilled for September 12, but the wind kept changing. Politicians and press were twice flown in to see it, and sent away disgruntled about a wasted journey. Once it was called off just an hour before the blast, when the wind switched and put the nearest city of Adelaide at risk of fallout.

While they waited, the indoctrinees helped position the dummies and target vehicles, along with military equipment such as mortars, machine guns, field telephones and radio aerials that would also be exposed to test their subsequent operability. They also mucked in with the building of a cinema in Maralinga. Finally, at 5pm on September 27, 1956, the weapon - the Hurricane redesign known as Red Beard - was detonated at the top of a 100ft tower. A report by Indoctrinee Force commander Brigadier Bethell described it as 'a most magnificent and awe-inspiring sight'. The next day, the *Adelaide Advertiser* front page carried three pictures of the mushroom cloud as it climbed into the sky. Under the headline 'Explosion Sets Desert Ablaze' it read: 'After 15 nerve-wracking days of postponement, Britain's sixth atomic weapon was exploded successfully at Maralinga at 5pm today. The fiery mushroom cloud which followed the explosion of the bomb from a tower surged over the proving grounds before the biggest audience ever to watch a British test. Towards sunset, a plane reconnaissance by pressmen revealed a grim harvest.' It reported how people heard the blast hundreds of miles away, it had set the desert brush alight, and the ground shock was registered on distant seismographs.

The *Canberra Times* carried a long description, which talked about 'red-hot fragments' of the weapon being 'scattered over the range', while tanks, trucks and buildings close by had burst into flames. It went on: 'The fireball, a white-hot mass, expanded into a gigantic bubble. All minerals and rock in the tower area were fused by the intense heat as the air heated to incandescence. A second or two after the explosion, the fireball shot upwards and as it lost its intensity it continued to expand with a brilliant orange glow… explosive gases swirled up, forming the head of a mushroom, and earth sucked up by the fireball formed the stem… a plume of radioactive dust fell from the cloud as it began to move up the mainland.' A foreign correspondent for *The Times* took a different tack, sniffing: 'The blast wave was disappointing, no stronger than the wind from London's Underground that strikes one in the street.' When journalists reported the wind had changed direction again, and fallout had drifted over the mainland, the Australian government accused them of misleading the public.

But yet again, a black mist was seen. A telegram sent on October 2, 1956 to the director of the Maralinga range from native patrol officer Walter MacDougall, who was camped on the extreme eastern edge of the range at Coober Pedy, said: 'Men employed at Ingomar who were camped 15 miles west of Mt Penrhyer [sic] bore on 27th September report that a very bleak [sic] cloud detached itself from main cloud and travelled northwards and then rejoined main cloud. This occurred about 1300 hours... At night when they were in bed particles of sandy dust were hitting their canvas camp sheets very similar to raindrops.'[42]

A day after the explosion, the safety committee told reporters the test had a complete 'bill of health', and that 'all dangerous fallout has been deposited and the remaining fallout is completely innocuous'. The statement said: 'There was no possible risk of danger or harm now or at any future time to any persons, stock or property.' Any radioactive rain might last for days, but would be 'quite harmless'.

But the Brits had promised no bombs on the mainland would be bigger than 10 kt, and this one was 15. Most of the indoctrinees had been on a hillside less than five miles from the explosion. Brigadier Bethell's later report said: 'This was a perfectly safe place and had been chosen in consultation with AWRE advisers.' A further three Australian officers were in shelters just under a mile from the blast. Afterwards, survey teams in Land Rovers drove in 'to ground zero to confirm that the site previously chosen… was clear of fallout'. How this was done is a mystery, as radioactive sand looks the same as the safer sort, and no alpha radiation checks were conducted. Bethell said that 'when such confirmation was forthcoming', the indoctrinees were ordered into the zone. Over the next two days, the troops were divided into 14 teams, each with a leader, a map, protective clothing and respirators, and taken on a two and a half hour tour to see damage to vehicles and dummies. On their return each showered, had radiation readings taken, and dosimeter badges recorded. Penney gave personal permission for the visits to be repeated for several days, and on subsequent trips 'only a small amount of protective clothing had to be worn', said Bethell. He said no man exceeded his maximum permissible dose. But none were being monitored for alpha radiation, and they were about to get even closer to it.

The next bomb, Marcoo, would be a ground blast in a concrete pit in the desert, intended to suck up more sand, and then drop it as fallout. It used the Blue Danube warhead, but with a low-yield core so that its dust would not be so deadly it could not be studied. Some of the indoctrinees were 'so enthused' by their first experience they volunteered for a second. Two dozen were ordered to dig a trench with overhead cover 2,000 yards from the bomb, 'to experience the ground shock'. Four were inside another Centurion tank, and another 80 or so were ordered to stand in the open, 3,200 yards out from the blast. These consisted of 38 British officers, three civilian defence scientists, 10 Australian officers, 17 Australian support staff, two New Zealand officers, another two health physics assistants, nine British military observers and the controlling Senior Trials Scientist. 'Health Control arrangements were as strict as those on Round 1,' crowed a report. 'All Indoctrinees were issued with new film badges, which they were required to wear at all times.'

The detonation happened on time at 4.30pm on October 4, with a yield of just 1.5 kt. It left a crater 160ft wide and 40ft deep, and the airborne cloud it created was tracked crossing the east coast of Australia 30 hours later.

A report written by a Major Janisch described how, on day three, 24 men were used to 'investigate the degree to which the clothing becomes contaminated'.[43] They were split into three groups of eight, wearing either battle dress, khakis, or 'AWRE protective clothing consisting of cotton overalls, short rubber boots and overshoes'. They all had respirators, and standard-issue 'combination underwear' and 'grey Army pattern socks'. They were taken to one of the nearby desert tracks 'at the point where the dose rate is highest' - around 10 mSv an hour, which was about five times what they would get as a background rate in the UK over the course of a year.

The first group sat on an open-sided three-ton lorry as it drove up and down the track for a mile at 20mph, taking turns so each spent an equal amount of time at the rear. The truck dragged behind it 'drafting sacks in order to raise dust from the ground'. The second marched 'in the normal manner' alongside the track for two miles, passing the lorry 'at frequent intervals'. And the final group of men leopard-crawled on their elbows 'in the accepted military manner', then for another mile marched 'through bushy areas to ensure that as much contamination as possible gets on their clothes'. They were 'brushing against trees, stirring up dust and crawling through the undergrowth for about 100 yards'.

But it was all for nothing - afterwards, the scientists found the dosimeter badges recorded too little to be any use. A report states: 'Contamination was so small that almost no information was obtainable. Again, all of the recorded [gamma] radiation doses… were comfortably within the upper limits anticipated in early calculations and well within the permissible levels.' This was poor science. It was known the badges recorded only gamma and beta radiation, which were the most powerful types, consisting of invisible beams like x-rays, but also quickest to decay. They did not measure alpha particles, the biggest constituent of fallout, and so weak that it can be blocked by a single sheet of paper. But once inside the body it cannot easily get out, and becomes so pernicious that it is likely to remain a risk for a much longer period. While it was possible to measure alpha using a Geiger counter, the sensor would need to be within millimetres of the particle to detect it.

There is no indication - aside from Dr Ellis' report during Mosaic - that alpha emissions were checked at any of the tests. In 2018, Parliament was told that: 'Alpha radiation was not routinely measured to monitor the environment because measuring beta and gamma radiation was more efficient.' Instead, scientists seemed to have decided that if the gamma readings were low, the alpha must be correspondingly minimal. It may be fine in theory but could never be verified if no-one bothered to check that the sums reflected what was, literally, on the ground. All the indoctrinees had were the badges - which were known to be affected by extremes of heat on the ground, and cold in the air. They were not always processed or stored properly, frequently lost, and the chemicals used to develop the film were kept in drums that sometimes spoiled. Simply put, two dozen men were covered in dust to see how radioactive it was, but without any means of measuring it.

Even those experiments that did not involve live subjects produced bad news. For the first two Buffalo shots, 'a standard man was constructed from mahogany wood'.[44] This 5ft 11ins dummy had dose badges attached in areas that corresponded with the location of human organs - the lung, spleen, testicles, intestine, and even - not standard service issue at the time - left and right ovaries. The aim was to see if the dose for different parts of the body matched up with that recorded by the badges, which were usually worn on the chest pocket. An AWRE report published six months later, in April 1957, states: 'The dummy was dressed in the standard protective clothing (underwear, boiler suit, socks, boots, gloves and a gas mask) and placed, after firing, in an area known to be contaminated with fission products, until it had received a dose of about 300 millirep [about 3mSv].' Despite the fact the dummy had not been marching or crawling as the indoctrinees had, nor drinking water and eating food, each 'organ' registered exposure. The average dose to the human body, according to the badges, ranged from 0.86 Roentgen in the spleen to 1.53 for the testicles, about a thousandth of the dose needed to kill a man outright. When the badges worked, they were still inaccurate. 'The intestine and testicles receive about 40% more dose than that recorded on the dosimeters normally worn,' said the study. Yet the badges were not changed, and no concern was shown for the men whose gonads were more exposed than the official record showed.

A week after Marcoo, Squadron Leader Ted Flavell of 49 Squadron became the first British pilot to drop a live nuclear bomb. A D-Day veteran who had flown dozens of missions across Nazi-occupied France and Norway to drop secret agents and supplies behind enemy lines, he also towed gliders across the Channel for the ill-fated attack on Arnhem. He and bomb aimer Flight Lieutenant Eric Stacey trained for the mission at Orford Ness on the Suffolk coast, the flat munitions range where 'bouncing bombs' were tested a decade earlier.

Kite was another low-yield version of Blue Danube. A 40 kt bomb had been downgraded to just 3 kt, in case the timers went wrong and it detonated on the ground. Flavell's log states: 'Conditions cloudy most of the way, and very bumpy while on approach.' Nevertheless they climbed to 30,000ft, lined up the target, and Stacey let the payload go at 3.27pm on October 11, 1956. The bomb exploded 490ft above ground and within 110 yards of the aiming point. The test was considered a near-perfect demonstration of what the RAF could achieve, and both Flavell and Stacey were awarded the Air Force Cross as a result.

The fourth and final blast, Breakaway, came 10 days later. This was a 10 kt version of Red Beard, atop a 100ft tower. Conditions had been perfect during the day, but the Australian government refused to countenance a firing on the Sabbath. For reasons of theology it was therefore detonated at five minutes past midnight on October 22 - technically a Monday, but only just.

The *Queensland Herald* reported the bomb was seen 600 miles away, with witnesses describing a 'brilliant, deep orange flash, with pale pink edges', 'very bright sheet lightning', and 'a flash of lightning spread across the horizon'. The newspaper claimed: 'Scientific measurements were made as before and the post-firing programme of tracking and checking the atomic cloud was in operation. Everything was proceeding according to prediction and normal safety precautions had been taken at the range.' As Penney left Australia a week later, he told Press there had been 'no possible danger' from the testing. 'They went off very much as we expected; in fact, they were completely successful,' he said.

But it transpired that at every single one of the Buffalo blasts, something had gone seriously wrong. Fallout was found across Australia, and in Maralinga Village where troops lived, in doses high enough to skew the results of scientists checking their samples. Each of the Buffalo bombs was detonated in violation of

The Kite blast was considered a near-perfect drop by the RAF, but exploding within 500ft of ground level meant the stem of the mushroom cloud met the ground, sucking up soil, irradiating it, and creating fallout.

the rules laid down by the Australians. One Tree contaminated Aboriginal lands, Marcoo was fired when rain was predicted, Kite took place under adverse weather conditions, and Breakaway dropped fallout on inhabited areas, overlapping that of previous weapons.

Radioactive iodine was found in the thyroids of cattle across the continent. The *Adelaide Advertiser* reported in 2001 that a sick rabbit had been captured and killed in 1956, and its carcass sent to the UK for tests. It was later found 'to be suffering from external radiation as well as ingestion trouble', with exposed teeth, its fur and skin around the mouth 'eaten away', and that it had gone blind in one eye as it had 'festered'. It had been found in the forward area, where indoctrinees had been sent.

There were multiple reports of Aboriginal people on the range, including large groups, and campfires. Soldiers who radioed these sightings in were greeted by hostility and disbelief. One was warned of the damage he was doing 'by finding Aboriginals where Aboriginals cannot be'. An aerial search was conducted only *after* the first and third bombs were detonated. When two patrol officers found a family on the range, they ordered them to leave and head to a further settlement. One offered a ride in his Land Rover, but was told off by his colleague for suggesting it, as indigenous people were not allowed in the vehicles. Instead the family was told to walk along the road - an illogical idea to those who walked between food and water. Nevertheless, the family did as they were told, and without sustenance on the white man's road, 'the mother, father, and brother perished'.[45]

Despite the risks, more dirty work was needed. The Kittens tests had helped redesign the triggers, but Penney needed to try the new version.

The first place suggested was an island called Foulness in the Thames estuary. It was then, and remains, both an important habitat for migratory birds, and a Ministry of Defence munitions range. But Parliament was queasy about Penney's experiments taking place so close to what they considered civilisation, and within a few months it was ruled out. A more remote site was considered at Wick in the far north of Scotland, where any contamination might blow out to sea, but wet weather was an issue, and again the Brits were sniffy about having it in their own back yard. Lieutenant Colonel Keith Stewart of the AWRE said it was dismissed out of hand. He said: 'I doubt if the people owning the estates in Scotland would look on that with very great favour. They are interested in pheasants and deer in Scotland.'[46] Admiral Patrick Brooking, then assistant deputy director at AWRE, let the cat out of the bag several years later when he wrote a memo to his boss, saying: 'I know you feel that radioactive contamination in the UK is politically impossible.'

There was no option but to return to Maralinga, which had the benefits of being ready to house troops and scientists, and offered the privacy needed for top-secret tests. Another benefit was that these non-nuclear experiments were not included in the written memorandum of understanding with Australia, which therefore could not veto them. When their government did ask about what was going on, the British refused to reveal what they were exploding. Titterton's safety committee declared it safe without knowing what was happening, and one Whitehall mandarin wrote a memo saying: 'I agree with Titterton that we should avoid formal communications on these contentious experiments and propose that we proceed without going through the normal channels.'[47]

The experiments were known as assessment tests or 'The Maralinga Experimental Programmes', but the name that stuck was the Minor Trials, ironic since they were perhaps more deadly than a full-scale nuclear explosion. It was imperative no-one think a single radioactive particle had gone astray. Finally, 24-hour guards and man-proof fences were installed in some parts of the range.

Kittens was restarted, to study the timing, intensity and ingredients needed to kickstart a nuclear blast. These 'initiators' had two ingredients - a radioactive one, and a toxic heavy metal, with detonations timed to the millisecond to mix, then compress the ingredients. The first five experiments had strewn about 750g of beryllium, 120kg of uranium, and 1.5g of polonium-210 across a small area. Polonium is so very radioactive that just half a gram of it can reach a temperature of 500 degrees Celsius. All the world's nuclear reactors produced no more than 100g in 2019, yet the British scientists were able to spray a hundredth of that in just a few months. There were dozens more of these blasts over half a decade.

In 1955, Operation Tims was launched, followed by Operation Rats a year later, both studying pressure waves inside the weapon. In Tims they tested different jackets around cores containing low-grade, natural uranium, and the reactions were measured minutely with telemetry and high-speed photography. In Rats, a more radioactive gamma source was added to the weapon, which meant in effect they could x-ray the blast as it happened. It used an isotope called scandium-46, which is often used today to trace liquids in oil pipelines, along with weapons-grade uranium-238, and a small amount of radioactive polonium. It is found in tiny amounts in the soil and air, but concentrates naturally in leaves of the tobacco plant. Scientists have speculated that lifelong heavy smoking can produce the same ill health as the fallout of a nuclear accident. Some Rats tests also used radioactive lead.

These smaller experiments were vital to the enterprise and ran until the 1960s, feeding information into the overall weapons programme. With no big bangs, flashes or clouds to disturb the public, the Minor Trials caused little concern at the time. Penney admitted to Australian newsmen only that 'tests would be conducted here periodically, depending on the speed with which Britain wanted to develop her atomic weapons programme'. But it was all a step towards the ultimate goal of a hydrogen bomb.

Within a decade of the Second World War, Britain had exploded nine atomic weapons and made them small enough to flick at an enemy. It still wasn't enough. The Americans and Soviets both had more powerful bombs, and the arms race was in overdrive. In March 1954, the US conducted the Castle Bravo test at Bikini Atoll in the Pacific. The cloud was four miles wide, and 22 miles high. It had a yield of 15 megatons - 30 times stronger than anything the Soviets had at the time, and double what even the Americans expected. Conducted at sea, the images it produced show a godlike power, with radioactive judgement raining down, without a single bullet being fired.

It also spread radiation across 250 miles of the Marshall Islands, and accidentally exposed the crew of a Japanese fishing vessel which wandered into harm's way. The *Daigo Fukuryū Maru* - Lucky Dragon No5 - was fishing for tuna outside the danger zone, but was coated in a fine, white powder, made of radioactive coral and sand. It took the crew six hours to retrieve their gear and process the catch, and they also scooped the ash into bags with their bare hands. It got in their ears, noses and mouths, and they had succumbed to radiation sickness on return to port. Their skin ulcerated, their white blood cell and sperm counts plummeted, and their gums bled.

The authorities ordered tests on any fish caught in a 2,000-mile radius around the test site, and thousands of samples were found to be irradiated. They gave a sample of the ash to Tokyo University, which found isotopes of strontium, caesium, and uranium.

One sailor was dead from complications of radiation within six months; others in the crew died years later from cancers of the lung, colon, stomach and liver, developed cirrhosis, stomach ulcers, and reported stillbirths. Japanese newspaper *Yomiuri Shimbun* reported 'Japanese fishermen encounter Bikini A-bomb explosion test. 23 men suffer from A-bomb disease', and within days the international press had picked it up. Stories of a 'death ash' and horrific injuries were picked up in Britain, particularly by servicemen's families. At the same time, more than 80 Pacific islanders were relocated. The Americans said 'their injuries have had a depressing effect on the morale of natives who have seen them'.

Some US co-operation was beginning to return, even though it was illegal. Only the British had planes capable of flying high enough to sample nuclear clouds this big, and so a deal was done - if the RAF loaned the Canberras, Penney could keep some of the samples. In greatest secrecy, crews of 1323 Flight and 540 Squadron were ordered to the US base of Kwajalein. It was codenamed Operation Bagpipes, and it was a messy affair. A report by Wing Commander Wally Kenyon, who led the team, says they lost one bomber on the way, along with its crew. Condensation had got into the radio compasses, and frozen at altitude. They were never found, but the search was hampered by the fact neither the British nor the Americans wanted to say where they were headed, or why.

A second bomber, tail number WH697, got lost and crash-landed on an atoll. The crew were rescued only after an islander paddled 100 miles to the nearest US coastguard station, with a note from the pilot asking for help. Patrick 'Pete' Peters, a young flying officer who had made it to the US base safely, was on the rescue team, and recalled how anxious the Americans were about maintaining secrecy by asking to remove all RAF markings from his aircraft. When he got to the downed plane, he saw the landing gear was bent and it would never take off. 'This damn thing was out in the open with RAF markings all over it, and had to be got rid of,' he told the *Sunday Mirror* in 2024. She was stripped of everything useful, including engines and rudder, before being towed out to sea. She stubbornly refused to be sunk, and was subjected to gunfire and then ramming before she succumbed to the waves. Then, said Kenyon, the planes which were still available to fly only got off the ground thanks to spare parts found in local workshops. He wrote that recorded doses of the crew were a fraction of permissible levels, but only because the radiation meters did not work, and that they did not even have enough rags to wipe the planes down, after they had been sprayed with detergent to remove radiation picked up in the cloud.[48]

At the end, Kenyon notes: 'There are two firm ends to a test programme of this scale. At one end there are great and powerful nations, rich in the products of the earth and of the mind; blessed with many skills and a restless energy to pursue ever deeper their inquisition of nature. But at the other end, sitting in the shade a few yards from my door, there is a little brown child whose wide eyes see none of these great things, but whose hair is gone and whose skin has fallen from a small

Flying Officer PHJ 'Pete' Peters took this photograph of the stricken Canberra bomber WH697 in February 1954, as she was towed out to sea from Ailingolapalap Atoll. (Courtesy of Pete Peters)

body which may one day grow cancerous.'[49] Although his report was declassified in 2010, a section on contamination levels will remain closed until 2037. By that point it will have been 82 years since the test, and everyone involved will be dead. Despite Kenyon's warnings, the Brits became more determined.

Atomic weapons are rated in kilotons, or how much energy they release compared to 1,000 tons of TNT. H-bombs are counted in megatons, each of which is 1,000 times stronger still. While an atom bomb can destroy a city centre, an H-bomb could take out 100 square miles. A planning document from the time, discussing how to respond to an H-bomb attack said, chillingly: 'Whether or not the enemy was able to repeat his initial assault, the devastation caused by a thermonuclear attack would be on such a scale that the UK could not be used as a main supply base... It is therefore no longer realistic to make plans for defence production in war, or for the build-up of industrial potential in peace, which are based on the assumption that this country could be used, after the outbreak, to support continued hostilities in a world war.'[50] Clearly such devastation would cost millions of lives with no way to save them, but the document makes it clear the public should not be told. 'The government, we suggest, should not seek to impress the public with the dangers... until they can tell them at the same time what measures of protection can be taken,' it said. The only defence was to be capable of doing the same to your enemy. Which is why the Cabinet, led by Churchill in the sunset of his political career, took the decision to build a H-bomb on July 27, 1954. It was announced publicly the following year.

Operation Grapple was what the entire nuclear testing programme had been building up to. Britain's atomic devices were to become triggers for fusion weapons with the potential to be 1,000 times more powerful than Hiroshima, and a million times more terrifying. The Chiefs of Staff were warned in December 1955 that: 'Political pressure both in this country and throughout the world against further megaton trials is increasing despite the firm stand taken by HMG in Parliament. It is possible therefore that after our first megaton trial series, further trials may become politically unacceptable. It is essential therefore to safeguard the future by obtaining the greatest possible amount of scientific knowledge and weapon design experience as the foundation of our megaton weapon development programme.'[51]

A week later, a second meeting revealed that although the scientists had almost no idea how to build an H-bomb, they knew enough to build 'an experimental assembly'. It would need to be 'arranged in great haste' and 'introduced by very careful publicity. Further tests might prove to be out of the question and we should therefore make the most of this opportunity'. Officials and scientists even decided to spin their plans when presenting them to government, warning 'it should be phrased in a manner calculated least to alarm ministers'.

Australia and New Zealand both refused to host a hydrogen bomb. Antarctica was considered. In the end, three islands in the middle of the Pacific, jointly claimed by the UK and US, were chosen, and the Americans agreed so long as they could monitor the whole affair. Sir Frederick Brundrett of the MoD said it was 'essential to obtain authority now' for the blasts, even if they were several years off. Penney said they could be delayed a little, as the missiles for which they wanted the warheads designed would not be ready for another decade. But the Foreign Office representative chipped in that the MRC report on genetic hazards, which was due to be published, 'might seize the imagination of the public'.

The memo warned: 'Our scientific knowledge of megaton warhead design is, at the present time, entirely unsupported by experimental evidence.' The plan was, therefore, to simply make a very big atom bomb and see if that worked. The planners reckoned it would be possible in two years, and the planned yield of up to seven megatons would add 'only 10% to the man-made atmosphere contamination'. They agreed the islands would be evacuated for each test, in case of accident. They did not consider that between tests there would be any risk at all.

But telling the world Britain intended to build an H-bomb would inform every foreign state of how advanced, and yet still vulnerable, the British were. A propaganda campaign was needed to calm the waters, so

War hero Air Vice Marshal Wilf Oulton was given command of the hydrogen bomb project in the Pacific. (Wikimedia Commons)

Penney gave a radio address to the Australian public. He also met with Dr Herbert Evatt, leader of the Opposition Labour Party in Australia. He told Penney: 'I don't agree with what you're doing, I think it's terrible.' Penney replied: 'I don't like it much, but I think we're going to stop war that way.'

Air Vice Marshal Wilfred Oulton was named task force commander for Operation Grapple. His father was a chemistry teacher, and as one of eight children young Wilfred needed a scholarship for university. He became squadron leader before war broke out in 1939, was mentioned three times in despatches, and honoured twice - once with the DFC for sinking three German u-boats in the Bay of Biscay, and later a DSO for protecting the mid-Atlantic gap with a squadron of Lancaster 'Flying Fortress' bombers. He was picked to be commander when the first choice, an admiral, refused the job, and because it was thought his experience in establishing a wartime bomber base at the Azores would be useful in militarising Christmas Island, a tiny coral atoll miles from anywhere.

Thirty years later he wrote a book about the mission, called *Christmas Island Cracker*[52] An obedient officer, he had followed instructions not to keep personal diaries at the time, and it was written from the best of his recollections. He remembered fears that Britain would no longer have a place in world politics, that the US would be ruthless to a weak ally, and that 'as for the Russians, the only thing they understand is a bigger axe than their own'. Christmas Island was chosen purely for its remoteness because 'if a nasty accident occurs, there'll be no significant risk to people downwind of the fallout and all the international uproar that would cause'. Being in the middle of the Pacific, it was also, coincidentally, the maximum possible safe distance from Whitehall.

The scientific director for the tests was Sir William Cook, a former civil servant and mathematician who'd designed arms for the UK government since 1928. He was Penney's deputy, and was frank with Oulton about their ignorance of the H-bomb. 'We shall be exploring an area which is unknown so far as we are concerned,' Oulton recalled him saying. 'We think we've got the right ideas - several of them - but we won't know until we test them and see if they work at all.'

Oulton asked: 'What sort of yield do you think you are going to get?'

Cook replied: 'God knows! Might not work at all... might be a fizzle of a hundred kilotons, or might be 10 megatons or more... we'll just have to wait and see.'

In June 1956, as *Diana* was preparing for her second trip through fallout, the British government went public. Eden had said six months earlier that he was open to a ban, but now he told Parliament: 'Her Majesty's Government have decided to carry out a limited number of nuclear test explosions in the megaton range. These will take place during the first half of 1957 in a remote part of the Pacific Ocean. The explosions will take place far from any inhabited islands and the tests will be so arranged as to avoid danger to persons or property. The tests will be high air bursts which will not involve heavy fall-out. All safety precautions will be taken in the light of our knowledge and of experience gained from the tests of other countries.'[53]

But Christmas Island *was* inhabited, by hundreds of people mostly descended from plantation workers and fishermen. First 'discovered' by James Cook on Christmas Eve 1778, it is often confused with another isle of the same name in the Indian Ocean. Today it is known as Kiritimati, the biggest island in the Kiribati republic, and as the Gilbertese islanders who live there pronounce t like an s, phonetically it still sounds like the name Cook gave it: Christmas.

The atoll had been occupied and abandoned many times due to drought and other issues, and was a military base for the US during the war. In the 1947 census the number of official residents was 47, but at the time of the tests the population was 229. For these bombs, though, they were expected to be out of harm's way. Christmas would be the task force base, with the bombs dropped 400 miles to the south east, over Malden Island - which, while it had once been home to Polynesians, was without fresh water, and definitely uninhabited. A third, Penrhyn Island, would be the base for weather monitoring and the American observers.

And so Britain launched a massive military operation in near-total secrecy.

Servicemen from all three armed forces, AWRE scientists, and even two ladies from the Women's Royal Voluntary Service to run a tuck shop were despatched to the other side of the globe. HMS *Diana* had dropped stores there on her way home in 1956, and the first troops were diverted from Korea to get started building the biggest outdoor laboratory in the world.

But the safety precautions Eden boasted of were not definitive. While 96% of servicemen at Operation Hurricane had dose badges, at the first three Grapple tests the figure was just 2%. Scientists were about to play with the power of the stars, with no idea if it would work, and little in the way of public monitoring. They were more worried about getting the weapons fired. H-bombs could theoretically minimise fallout, and therefore the public protests, so long as they were detonated high enough above the ground that they didn't suck too much material up - about 8,000ft. But to fire a weapon that big and give the plane time to get away before it detonated, it needed to be dropped from about 45,000ft - 8.5miles up, higher than the cruising altitude of most modern aeroplanes. It required a bomber, and crew, that could operate in the stratosphere, where the weather, temperature and risks are very different.

Luckily the RAF had been thinking about this since the end of the war, and in 1955 it had taken delivery of its first Vickers Valiant bombers, each with four Rolls Royce engines buried inside the wing to minimise air resistance. They were designed specifically to drop nuclear weapons from a great height, and after their first trial run at Maralinga were ready to be unleashed. The other problem was that dropping bombs from such a height was often inaccurate. When the Americans tried it in 1956, they were four miles off-target. Penney wanted to get within 300 yards.

First Oulton and his task force needed to get the island ready. The wartime landing strip was resurfaced, lengthened and widened to take the Valiants. Bulldozers and fire engines were shipped from Britain as Royal Engineers built water tanks, distillation systems and roads with help from the islanders. Vast areas of scrub were cleared, buildings thrown up, generators shipped in, while army signallers built communication systems and radio masts. The RAF provided aircraft and meteorologists to produce weather forecasts for vast swathes of ocean, technicians modified fuselage valves to collect radioactive samples, and ground crew set up decontamination tents and showers. Barrack tents, huts, fuel storage and pumps were all built, along with mess halls and lead-lined laboratories for scientists to put the finishing touches to the weapons, and do their first measurements of the effects. There was even a post office, a butcher's, and a bakery.

Public opposition was growing, especially after reports - quickly refuted, but nevertheless worrying - of radiation going astray. Perth had seen a series of protest marches, with placards held aloft by people claiming that civilians were 'guinea pigs'. So at the end of 1956, in response to an official request for 'innocuous' stories featuring military personnel, troops were given 2,000 donated toys and told to write out festive tags for each of them. They were handed out in UK hospitals on Christmas Day that year, with accompanying positive write-ups. The aim was 'to discourage speculative stories arising from the press possibly contacting personnel returning from Christmas Island', by giving journalists

RAF navigator Dennis Hobbs took this photograph of the docks at Port London on Christmas Island from a Handley Page Hastings transport plane in August, 1956. (Wikimedia Commons)

something else to write about instead: a Christmas present from Our Boys at Christmas Island.

The foreign press was not so obliging. In April 1957, the *Fiji Times* wrote: ‘Nobody knows how many people will die or how many children will be born mentally or physically deformed because of atomic or hydrogen bomb tests, past or future… The free nations should seek foreign agreement with Russia to curtail or suspend completely all tests until their effects on the future of mankind can be more accurately assessed. To continue with indiscriminate and unrestricted tests in the present state of uncertain knowledge will be irresponsible folly indeed.’

At the start of 1957, islanders began to be evacuated. Around 100 were taken off by boat in the first couple of months, with another 29 women, 56 children, and 31 men leaving by aircraft in March. Thirteen men remained on the island to help the task force, while Royal Navy warships, with RAF planes overhead, began patrolling the testing grounds to keep onlookers away. At home, politicians were agitating against the bombs. The *Mirror* reported on May 1, 1957, that Opposition frontbencher Nye Bevan said the US and USSR were ‘not content at being able to kill the human race in three days’, so ‘they go on to see if they can do it in half an hour’.

In order to limit the tests, scientists were experimenting with weapon design at the same time. The plan was for three bombs, each intended to be a megaton device, but with slightly different delivery systems. The first, codenamed Short Granite, was too big to fit on a warhead but would test their theories. It would be followed by Orange Herald, a boosted fission weapon, but small enough to fit on a missile. The last would be Green Granite, a true thermonuclear weapon again carried by missile. A fourth device, an interim bomb called Green Bamboo, was to be included if one of the others didn’t work.

The final components for Short Granite were three days late in arriving, but after a dress rehearsal on May 11, 1957, it was decided the weather would be favourable four days later. The bomb was assembled under the direction of John Challens, a AWRE scientist who had designed the electrical circuits for Hurricane. He had been one of two who conducted last-minute checks on the device before leaving the *Plym*, and by 1957, had risen to be scientific director of the Christmas Island tests. Like many other mathematicians who graduated on the cusp of the Second World War, he had been recruited into the MoD armaments division. Penney invited him to join the team at Fort Halstead, working on detonators that could fire a millionth of a second apart. Over the next few years, he had to repeatedly refine the circuits to include different timers accurate enough to trigger a double-detonation fusion bomb. It turned out his designs were far in advance of anything the Americans had come up with.

Official observers from allies were flown in, and the entire task force gathered at Malden Island, inside a naval no-go zone. Shackletons patrolled the air while scientists, observers, and troops were aboard the aircraft carrier HMS *Warrior*, acting as the control ship, *Narvik* which was the base for scientists, *Alert* and *Messina*. Wing Commander Kenneth Hubbard, commander of 49 Squadron, took off in his Valiant XD818 at 9am on May 15. At 10.38am he announced ‘bomb gone’, and banked steeply to accelerate away.

Nick Wilson was a 19-year-old National Serviceman when he posed for this photo, on the right, with a friend in the navy photographic unit. In letters home he described 'the most amazing sight I am ever likely to see' and said the fireball 'looked like a very ripe, dark red apple with ice cream being poured over the top of it'. He took his own life 30 years later. (Courtesy of the Wilson family)

A news sheet produced for troops on the island announced that Britain had emerged from the blast 'as a top-ranking power of this nuclear age' and 'no human eye could survive the hellish glare of white-hot air brought to incandescence by the fantastic heat'. It reported how, within an hour of the blast, scientists were flown ashore from the *Narvik* to take readings. In the air, Canberra bombers began flying sampling missions. '*Warrior*'s helicopters flew towards the target area and took radiation readings from sea-level up to various heights,' it said. 'Ships sampled sea water at their locations.'

But it was a dud. The Brits did better than the Americans, but not close enough, with the bomb 418 yards off-target. Grapple 1 had 'fizzled', with a yield of 300 kilotons - massive compared to those which had gone before, but only a third of the power needed to demonstrate fusion. Worse, the Americans knew.

General Richard Coiner, the official observer from the US Air Force, remarked to Air Vice Marshal Ginger Weir: 'Congratulations on your wonderful organisation, Ginger, and how wise you were to begin with a small yield and work up.'[54] The Brits had no choice but to take it on the chin. Oulton bracingly called it 'an extremely good start' and Penney went back to the drawing board. It was too late to change the second blast, but he scrapped plans for the third and rebuilt it. He added uranium, made the jacket out of aluminium, and called it Purple Granite.

Although the bomb had been a disaster, there was no need to let everyone know. In London, the press reported it was an H-bomb. The PM, Harold Macmillan, told the Commons everything had gone to plan, and the testing would continue. He claimed Britain was 'in a very much better bargaining position' to persuade others to disarm. He added: 'Having made it, we had to test it. I can see the argument for not having the bomb at all - although I disagree with it - but I can see no sense in going to the trouble of making the bomb, and then stopping short and not knowing how it works.' He didn't admit that his government still had no such weapon.

All hopes were now pinned to Grapple 2, but that did not go smoothly either. The bomb components had arrived two days before the first blast, but building it took a fortnight. Fijian naval reservists were added to the crew on *Warrior*, and a press corps was flown out to watch what they thought would be a second H-bomb, and the scientists hoped would be their first. The *Mirror's* legendary columnist Cassandra, real name William Connor, was among them. As the country's most popular columnist, he was in a position to send a telegram to the paper's financial officer before he left, saying: 'I am obliged to go to the South Pacific on top secret mission. Kindly arrange accommodation at the Honolulu Hilton for an indefinite period.' Before leaving for Christmas Island, he also asked them to send a tailor to fit him with tropical whites, because 'I am to have dinner with the captain of the flagship'.[55] By the time he was in situ, properly attired and pen poised, he had to send a third cable: 'Boom! Postponed.'

Orange Herald was finally dropped at 10.44am on May 31. Alan Pringle was the 22-year-old co-pilot, and told the *Guardian* in 2018 how he and Squadron Leader Dave Roberts flew in total darkness, relying on instruments with metal visors over the windows to protect their sight, with just a sliver of welder's glass through which to note the explosion. 'After 50 minutes of flying we reached the rendezvous and the bomb aimer, one of five crew, went to lie down on a mattress in the well in the belly of the fuselage; from that position, he looked straight ahead through the bomb sight at the aiming point. We then made a practice run – up to the release point, then a sharp turn to get away,' he recalled. The bomb aimer flicked off the safety, and said 'steady, steady, steady' so the pilots held their course. He pressed the button, announced 'bomb's gone', and the plane lifted noticeably as the 12,500 lb device dropped away.

Then they hit a snag. 'Roberts rolled the aircraft into a steeply banked turn to get away from the upward rush of the exploding nuclear bomb. If we had carried on in a straight line, the fireball would have come up under the aircraft and that would have been it,' Pringle, then aged 85, told the newspaper. 'The trouble was the accelerometer wasn't showing any G-force at all. I told the captain 'More G!' but as he pulled back on the stick we started to judder; the aircraft was stalling. We had been turning, we just hadn't felt the added G. We realised the accelerometer wasn't working, and the overcorrection sent the bomb aimer – who was making for his seat – crashing back into the well... if we didn't turn, we wouldn't escape.'

Luckily Roberts got the plane back under control and the weapon exploded 53 seconds after release. Pringle recalled: 'Three minutes later, we felt a biggish bump as the shockwave hit the aircraft, then a smaller one a second later. We saw the sky lit up; 10,000ft above our eye level was a writhing molten mass. I was in a state of awe.'

Those watching below had no idea of the drama, and were gazing at an unfolding explosion several miles high. But this bomb hadn't worked properly either. At 720 kilotons it was the biggest Penney had ever achieved, but it still wasn't fusion. The ingredients hadn't been mixed right. The Press corps were gathered on HMS *Alert*, 35 miles off the coast of Malden, and they'd seen what to them looked like a very big bomb. When Cook informed them solemnly that it had been yet another successful thermonuclear device, they had no reason to doubt him.

Connor watched awestruck, and then cabled his despatch back to London. On June 3, 1957, the *Mirror* front page carried a government-released picture taken from a RAF plane circling the mushroom cloud, describing it as 'Britain's second and mightiest H-bomb exploding over the Pacific'. It included the claim this was a five megaton device, and repeated its earlier call for testing to halt now that 'Britain has proved to the world that we, too, possess this horrible weapon'. Inside, the newspaper carried a cartoon of a tiny Macmillan, given stilts so he could talk to the US and Soviet premiers looming over him, saying: 'Now we can talk as equals.'

Connor wrote: 'It was a dress rehearsal for the death of the world. Standing on the rolling deck of HMS *Alert* and clad in white protective clothing with hoods and goggles we, the observers, looked like grotesque mourners… We were 35 miles from where The Beast was due to explode after being spewed out of the bomber - quite near enough in view of the fact that the power of the bomb was equal to several million tons of TNT. I waited with feelings of excitement, awe and a faint sense of horror.' He said the countdown sounded like 'footsteps that lead to the execution shed', and described turning his back to the blast, covering his eyes, and waiting for the flash to pass before turning around to watch the cloud develop.

'Through closed eyes, through dark glasses and with my hands still covering my face, I saw the flash. Brighter than the sun, hotter than the sun, and ripped out of the secrets of the heart of the universe,' he wrote, and then he turned. 'AND THERE IT HUNG BEFORE US, A BOILING RED AND YELLOW SUN LOW ABOVE THE HORIZON. IT WAS AN OIL PAINTING FROM HELL, BEAUTIFUL AND DREADFUL, MAGNIFICENT AND EVIL.' He gave a vivid description of the fireball, telling how it turned from gold to orange, grey, and a 'light muddy purple' before becoming 'a bloated, top-heavy Christmas pudding, with a greyish, whitish sauce streaming out of the top and spilling down at the sides like a filthy lava.' He saw shockwaves shimmering out of the cloud, and heard grunts as the pressure passed over the deck. 'The men around me were too quiet, and in a blasphemous

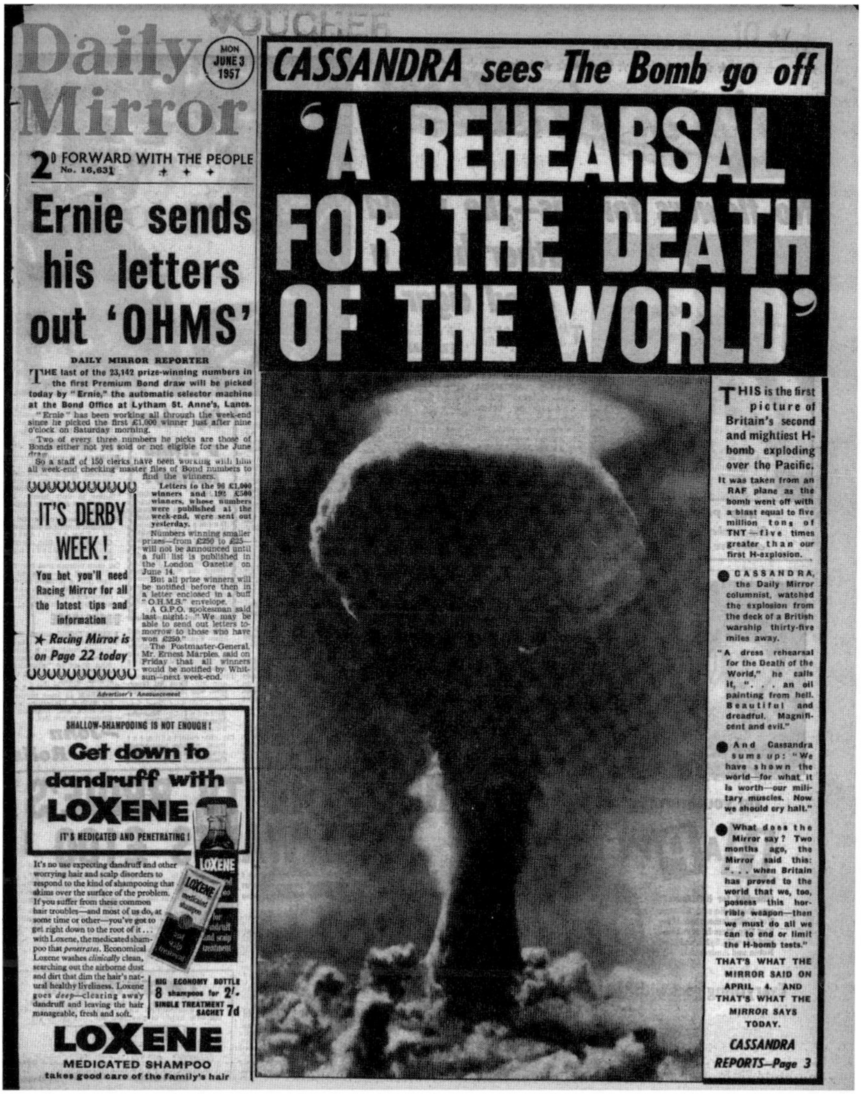

Daily Mirror

MON JUNE 3 1957

2d FORWARD WITH THE PEOPLE

No. 16,631

CASSANDRA sees The Bomb go off

'A REHEARSAL FOR THE DEATH OF THE WORLD'

THIS is the first picture of Britain's second and mightiest H-bomb exploding over the Pacific.

It was taken from an RAF plane as the bomb went off with a blast equal to five million tons of TNT—five times greater than our first H-explosion.

● CASSANDRA, the Daily Mirror columnist, watched the explosion from the deck of a British warship thirty-five miles away.

"A dress rehearsal for the Death of the World," he calls it, ". . . an oil painting from hell. Beautiful and dreadful. Magnificent and evil."

● And Cassandra sums up: "We have shown the world—for what it is worth—our military muscles. Now we should cry halt."

● What does the Mirror say? Two months ago, the Mirror said this: ". . . when Britain has proved to the world that we, too, possess this horrible weapon—then we must do all we can to end or limit the H-bomb tests."

THAT'S WHAT THE MIRROR SAID ON APRIL 4. AND THAT'S WHAT THE MIRROR SAYS TODAY.

CASSANDRA REPORTS—Page 3

Ernie sends his letters out 'OHMS'

DAILY MIRROR REPORTER

THE last of the 23,142 prize-winning numbers in the first Premium Bond draw will be picked today by "Ernie," the automatic selector machine at the Bond Office at Lytham St. Anne's, Lancs.

"Ernie" has been working all through the week-end since he picked the first £1,000 winner just after nine o'clock on Saturday morning.

Two of every three numbers he picks are those of Bonds either not yet sold or not eligible for the June draw.

So a staff of 150 clerks have been working with him all week-end checking master files of Bond numbers to find the winners.

Letters to the 96 £1,000 winners and 192 £500 winners, whose numbers were published at the week-end, were sent out yesterday.

Numbers winning smaller prizes—from £250 to £25—will not be announced until a full list is published in the London Gazette on June 14.

But all prize winners will be notified before then in a letter enclosed in a buff "O.H.M.S." envelope.

A G.P.O. spokesman said last night: "We may be able to send out letters to-morrow to those who have won £250."

The Postmaster-General, Mr. Ernest Marples, said on Friday that all winners would be notified by Whitsun—next week-end.

IT'S DERBY WEEK!

You bet you'll need Racing Mirror for all the latest tips and information

★ Racing Mirror is on Page 22 today

The *Daily Mirror* front page of June 3, 1957, had no way of knowing that Britain's reported 'second H-bomb', while 48 times bigger than the blast at Hiroshima, was actually another dud. (Reproduced with the permission of Reach plc)

way it reminded me of the silence... of Armistice Day,' he said. 'The flash, crash and roar of the hydrogen bomb set off in the most remote and desolate part of the world is a source of wonderment and indeed pride, to some... but when released over cities where it would obliterate millions of men, women and children in a trice, it is a wicked, an evil thing.'

Connor was sceptical about official claims these weapons were safe. He reported how Cook was 'almost enthusiastic' about the small odds of anyone suffering fallout from the blast, and mocked him for making it sound like 'an almost... hygienic weapon'. He pointed out Cook addressed journalists while 'wrapped from head to foot in protective clothing and wearing a device to detect

Bill Connor sits on the deck of HMS *Alert*, his portable typewriter by his side, as he waits for the blast. (Courtesy of the Connor family)

excessive radiation'. Connor died just 10 years after the blast, at the age of only 57. He had developed diabetes, a condition also seen in other survivors, and which his widow Isobella later blamed on exposure to radiation. It led to a fall and head injury, which ultimately claimed his life, a year after he had been knighted by the Queen for services to journalism.

When retired Squadron Leader Stephen Pooley died in 1996 from leukaemia, aged 82, his family also thought it could be linked to Grapple. The coroner ordered an inquest, and called as witness an MoD civil servant who swore under oath that Sqn Ldr Pooley had been too far away from the explosions to experience any ill-effects, and there was 'no proof' he had been at any risk. But in 2007 a declassified report written by Pooley showed it to be a deliberate lie. Not only had he escorted 12 barrels of fissile material for Orange Herald from Aldermaston to Christmas Island, but he had watched the weapon being built, and needed repeated decontamination after observing the removal of air samples from its mushroom cloud. He could not have been much closer.

Pooley's report[56] says he was on Christmas Island for almost a fortnight. His tasks were to monitor the fissile material, in this case enriched uranium, during the journey; help with cloud sampling preparations; observe the preparation of the weapon and 'if possible, to witness a burst'. Pooley said the aircraft were covered in white paint before each mission which, when washed off later with a bicarbonate solution, took most of the radiation with it - except for the hatches, which remained radioactive. The water, containing the contamination, 'is allowed to percolate away through the coral sand'. He was able to view the inside of Orange Herald as it was constructed, and described the arrangements of fuses and triggers, as well as how the bomb was photographed with a camera beneath the aircrew's feet as it dropped. At the same time, radio signals from the nose cone of the bomb transmitted information to the ground.

The report describes how the sampling planes had filter papers placed inside ducts in the wing tips, electronically operated from inside the cockpit. Another collector was on the top of the plane, in a 'periscope-like dorsal mounting', with particles entering via vertical slits to collect on absorbent paper. A rubber gas bag was carried in the belly of the aircraft, and once in the cloud the bomb bay doors would open and allow it to collect gas to a defined pressure before it was closed off. Finally, a smaller version of the wing tip collector was attached to the outside of the bomb bay door.

At the time of the blast, he said he was occupied 'on other duties' and did not even notice the flash. When the planes returned, he saw samples collected by scientists wearing full protective gear, taken behind a lead-brick wall and removed with tongs before being assessed, then packed into sealed cans inside lead-lined boxes to be flown home. The contents of the gas bag were decanted into a gas bottle.

The report has some glaring inaccuracies. It claims that not only did sampling crews take part in only one such mission each, when in reality they did repeated flights, but also that when the Orange Herald bomb was dropped the Valiant flew smoothly away. Pooley also wrote that it was a one megaton device. Clearly, the government propaganda was also being dripped into the ears of its own officers.

But most crucially for his family, he wrote: 'The most striking feature of the sampling procedure was the remarkably penetrating quality of the radioactivity. It was quite impossible to confine it. In fact the sampling tent was soon so active that the sample had to be taken outside for radiation counts. After four showers and a hair cut I was still above the permissible level of activity, which did not fall to normal until the following day.'

Many years later, Sqn Ldr Pooley's son-in-law had married the mother of a young girl called Sarah, who grew up to marry the Prime Minister of the day, Gordon Brown. Sarah Macaulay, as she used to be, had played on the Pooleys' lawn on family visits to Yeovil, Somerset, and called the retired airman 'Grandpa'. Her step-brothers and sisters were his grandchildren. Finally, it seemed that the impact of testing had arrived at Downing Street.

At the inquest, Pooley's widow Elspeth said he had never mentioned the tests and was healthy until the leukaemia struck him down. The coroner said the weapons trials 'had a Heath Robinson flavour', but with the MoD witness insisting there was no proof of exposure, and the report still classified, he recorded an open verdict.

Ken Moses was a lance corporal in the Royal Electrical and Mechanical Engineers. He joined 220 sailors to witness the first three Grapple tests from *Warrior*'s deck in 1957. Between blasts, he spent seven months living in a workshop on the western tip of Malden, a few miles from Ground Zero. His unit would travel across the 15-mile wide island, repairing and servicing kit and equipment and recovering broken-down vehicles. He said there was just one tree, along with two large and three small pigs, which avoided them. There was a small graveyard for victims of drowning, two stone huts used by people who came to harvest guano, and the stern of a shipwreck. While there, he was used in propaganda photos sent home to his local paper the *Stoke Sentinel*, along with others from his unit, about how they were enjoying the festive season. It called them 'bronzed soldiers from North Staffordshire in a tropical setting... united in sentiment with the late Christmas shoppers in the murky centre of Hanley'. In 1957, he was given a commendation from Oulton for his 'devotion to duty'. But by 1959, Ken's medical records show he had begun bleeding from his gastro-intestinal tract.

Health physics records he obtained from the AWRE show Ken's name among the 2% of personnel to get a film badge. Although no dose is logged for him, on the day of Short Granite one comrade received a dose equivalent to six months' worth of background radiation in the UK. For another serviceman, over the course of May, a figure equalling 10 years' worth of radiation is recorded. Yet in 2008, the then-Veterans Minister declared none of those who went ashore had any appreciable dose measured on their badges.

Ken recalled being sent back on to Malden the day after each blast. He said: 'When we returned after the explosion the scrub was smouldering. There were dead birds everywhere. The thing I noticed most was the silence, not a sound was coming from anywhere. The roofs of our building were collapsed inside, the boat shed roof was lying across the three boats, the urinal had gone. The only tree was withered and when we went across the island later an incinerated pig was found, with no sign of the other four. And the stern of the wreck had gone.'

In war pension applications, Ken's subsequent anaemia was attributed to iron deficiency rather than radiation, and he was denied a payout. In 2009, after a minister insisted the weapons had been dropped three miles out to sea, a TV documentary broadcast footage of them detonating over Malden. Ken found in his own papers a clear photograph of the target - a three-sided, concrete triangle 50 yards across, and sent it to the minister. But he was told: 'The AWRE records show that Malden was not a target, but an aiming point... the coastline of its southern tip was used as a reference.' There was no explanation of how aiming at Malden would guide a bomb drop three miles away, nor why TV footage showed something different. 'I know the truth, I was there,' said Ken. 'There is a culture of deceit.'

Behind the scenes, Cook and Penney were desperate for the third and final bomb to get it right. But when the rebuilt Purple Granite was exploded on June 19, it yielded a mere 200 kilotons - the dampest squib of all. Britain had spent a lot of money, expended a lot of energy, and made a very public fuss, about a bomb it didn't have. The Americans and Soviets knew they didn't have it, and public outrage was growing. Parliament ordered sites across Britain be examined for 'H-dust' after man-made isotope strontium-90 was found in the bones of sheep grazing on contaminated grass in Wales, Durham, Dartmoor and Lancashire. With two global superpowers already armed to the teeth, the rest of the world was clamouring for a ban on testing. If Britain was to find out how God made the stars, it had to do it fast.

Penney went back to the drawing board, and the Australian wilderness. He needed to know more about what happened in the first milliseconds of a detonation. Influence over these first moments could enable them to specify its power - and by extension, the power of Britain on the world stage. But the first plan, codenamed Operation Volcano and featuring six lightweight devices designed for missile warheads, held aloft by balloons and situated on towers at Maralinga so they could be detonated in mid-air, was stillborn. 'Balloon bursts' were more practical because they involved less construction, allowed the same area to be reused, and theoretically limited the amount of dust and sand sucked up. But the Australians were horrified at the idea of an untethered atom bomb drifting who-knows-where.

Polls showed just over a third of Australians supported a fresh series of tests, the reasons for which the British saw no need to discuss. The fiery codename, officials warned, could 'give rise to unfavourable public and press comment'. On top of this, the total predicted yield was up to 80 kilotons - more than double the combined power of Buffalo. To allay fears, Volcano was rebranded as Antler, which was altogether more graceful but still awash with machismo. Penney undertook a series of radio interviews about the peaceful benefits of nuclear power, and it was decided there would be only three blasts - two on towers, and just one tethered to a balloon. If it escaped, the Australian air force would shoot it down.

There was so much secrecy attached to the trials at this point that the tests were in danger of being called off. Concerns about widespread contamination meant the local authorities had to be seen to put their foot down. The first proposal by the British about five more tower tests was considered so 'general and unspecific' the Australians refused consent, pending more information. The Menzies government was 'aware that in the prevailing climate of public opinion any failure to assure itself of the exact nature of the proposed tests could be electorally damaging'.[57]

As time wore on and agreement was still withheld, Alec Douglas-Home, the UK's Secretary of State for Commonwealth Relations, was tasked with bringing the colonials into line and wrote to Menzies with further details. He said the yields would vary, with the biggest being similar to that of Mosaic II, although at the time the figure was still a secret. Three would be on towers and three held aloft by balloons, which would minimise the possibility of soil and dust being sucked into the cloud, irradiated, and deposited as fallout.

In the end, just three devices were fired. All were lightweight missile warheads, with a core of plutonium-239. The first tower bomb, Tadje, was exploded on September 14, 1957, with a tiny yield of just a kiloton. Biak, six times more powerful, was detonated on another tower less than a fortnight later. And the final blast, the 25 kiloton Taranaki, was exploded 1,000 feet up from under a balloon. The RAF had the most skin in the game - 31 aircraft and 700 men, 70 of them guarding the balloon. They were backed up with 450 Australian troops. The 170 AWRE scientists were supported by 46 Commonwealth boffins. Just 48 per cent of those present had dose badges.

Parliamentarians, international observers, and 20 members of the media were invited to watch. The British were so confident Antler would work that they drafted a press release, three days before the first blast, declaring it 'highly successful' and that 'cloud behaviour was accurately forecast'. In fact, the opposite came to pass. Tadje, which had been delayed for two days by bad weather, included a new and highly-toxic element - radioactive cobalt-60. Not known in nature, it is produced only in nuclear reactors, and pellets of the material were put inside the bomb as a way of tracing its yield. But no-one had told the Australians, and when lumps of cobalt-60 were found spread across the range it led to rumours that the British were developing a cobalt bomb - a horrifying device designed to render areas uninhabitable for long periods. It had long been feared as a successor to the H-bomb. Penney decided the cobalt didn't help, and it was not used again. It did, though, help the cloud samplers trace the fallout blowing north when it had been expected to go east, and detected it at Alice Springs, Melbourne, and Adelaide. Biak, which exploded 11 days later, had a cloud that rose twice as high as expected, and drifted all the way to the east coast, to Queensland and Cairns.

An air operations report shows the plan was for air crews to fly repeatedly through the cloud until they accumulated 2.7 Roentgen. In the event, it says that on entering the cloud at 19,000 ft just 30 minutes after the Taranaki explosion, 'the primary sampler recorded a cumulative dose of 10R within a few seconds of entering the cloud, and although he then left the cloud by the shortest route and returned direct to base his final cumulative dose was over 12R.'

The report concluded that the best air crews to use in such sampling missions were those who'd done it before, saying 'other factors being equal, crews with previous experience in the role produce the most reliable results... vast quantities of money and the best scientific brains are expended in order to advance our atomic knowledge, and we cannot afford to waste these resources... to produce such results experienced personnel must be used'. Ten of the six 'sniff' plane crews had taken part in previous tests, which meant they had already been exposed repeatedly. The authors of the report, circulated to the Air Ministry, Bomber Command, and AWRE, were keen to expose as many troops as they could to radioactivity. 'It is of vital importance to the RAF that as many personnel as possible be given practical experience of atomic weapons and the precautions necessary in the conduct of air operations in areas of atomic weapon activity. There is no substitute for practical experience in this sphere,' it says. 'Atomic trials such as Antler present an excellent opportunity of building up units in the front line capable of operating in an atomic war.' It suggested switching out 50 per cent of the crews every time, so that experience could be balanced out with fresh faces at every test.[58]

John Brothers died completely helpless, unable to talk, eat or fend for himself. As a navigator in 76 Squadron his job had been to track the clouds of Mosaic, Buffalo, Antler and, later, Tims and Grapple. A report written later by his pilot, Flight Lieutenant John Spatcher, said they were repeatedly contaminated. After flying through the Breakaway cloud in 1956, he and John 'were found to have a very high level of radioactive iodine and we had to stay at Maralinga for daily thyroid tests'. They had a dose equivalent to 17 years of background radiation in a little over an hour. At Antler, they flew for three and a half hours through the Biak cloud, but 'no radiation records exist'. Worse was to come at Taranaki. Spatcher wrote: 'Shortly after entering the cloud we became enveloped in a very thick black swirling patch and our radiation detection instruments showed maximum readings... our personal dosimeters, which showed the cumulative dose, were all at maximum after leaving the cloud.' On this flight alone, both

'Cloud flyer' John Brothers was a navigator in 76 Squadron ordered to fly through and track fallout clouds. He was unable to have children, and died from multiple tumours. (Courtesy of Wendy Brothers)

men recorded a dose of 6620 millirem - equivalent to about 33 years' worth of background radiation.

Spatcher added: 'After landing and entering decontamination, I was alarmed to see all the ground staff and scientists - which normally attended the aircraft and assisted us out of the cockpit - run away, and we were left by ourselves for between 5 and 10 minutes. When the door was opened and the edges sealed we were encouraged to depart the area as fast as possible.' The boffins did not approach the plane to collect the sample filters either, he said. 'We were told that the aircraft was so radioactive that it had triggered many of the warning alarms in the centre, and the personnel had departed to get advice on what to do. We were told that it was many hours later before the filters and other instruments were removed for analysis as it was considered too dangerous to approach the aircraft immediately after landing. I later received a letter and a photograph of the explosion from the Task Force Commander congratulating the crew on obtaining a "record sample".'[59]

Spatcher received the Air Force Cross in the 1958 New Year Honours, but John had a different outcome. Forty years later, after years of infertility, doctors found cancer in his throat and a tumour in his groin. He was 64. Two years on they found five more tumours in his brain, and John decided against chemotherapy. His widow Wendy, of Cranwell, Lincolnshire, said: 'It was horrendous – he was helpless as a baby. He ended up not knowing where he was or what was happening. To be honest I was glad for him when he died because it was so awful for him.'

The couple married 10 years after the trials. Wendy said: 'I already had one child from my first marriage, so we knew I was fine. But while with John I could conceive, I couldn't carry any children to term. You try to block it out so I'm not sure, but I think in all I lost six babies through miscarriage. We went to lots of doctors, and they said it was something to do with John but they didn't know what.' She added: 'I once asked John after he became ill in 1997 whether his service could have affected him in any way, bearing in mind the miscarriages. He said that as far as he was concerned his cockpit was sealed and the RAF would have looked after his safety.'

After his death Wendy demanded John's medical records to apply for a war pension, and found his fertility problems, as well as his monthly health checks, had been removed. Wendy died in 2021, still mourning John and their babies.

Taranaki was the blast everyone at Antler remembers. It was the biggest weapon ever detonated in mainland Australia. Three balloons were required to hold the bomb aloft, and a total of 12 were shipped to Maralinga. They were inflated with hydrogen, and moored out in the open 20ft up, rather than more safely in a hangar, which would be expensive to build. The air operations report says that 10 days before the Antler tests began, a storm hit the base, with catastrophic effects.

'The NCO in charge of the night handling party reported that one of the three balloons had sustained damage to a stabiliser... the balloon's behaviour becomes unpredictable and it is extremely dangerous to handle as it may drive to the ground on top of the handling party. The wind speed at ground level was 55mph and very turbulent at that time so an attempt was made to let out the mooring strops to their full extent in the hope that conditions would be less turbulent at 380ft. Before this could be done, the damaged balloon had been destroyed by fire after striking the ground, and a second balloon had become unmanageable due to damaged stabilisers and eventually destroyed itself. The third balloon was seen to burst into flames in the air', it reported.

Troops wore shorts and boots as they toiled in 40 degree heat, building towers, roads, and bunkers. The range had already seen four A-bombs and dozens of smaller devices, and veterans recall everything was coated in a thin film of sand. Dust blew everywhere, and scientists had to order troops to sweep it out of the areas where they were trying to work, because it was radioactive enough to interfere with the readings.

Brian Tomlinson was at Maralinga for a year during Operation Antler, and was left with bleeding ulcers on his palms for two decades. Then a Royal Engineer corporal aged 20, he was put in a 40-strong unit with Australian soldiers. While top brass and scientists were housed at Maralinga Village, Brian and his comrades camped under canvas deep inside the testing grounds, nine miles from Ground Zero.

'Nobody told us what it was all about, or checked us for radiation, but every morning we went into the forward area. The only barrier was this simple wooden post across the road. We'd roll up, they'd lift it, and in we went. Same when we came back out. Nobody checked us, did readings, or even washed the trucks down. We had pneumatic drills, and had to blast down through the soil. There was about 12 inches of earth, red dust, and below that was rock,' he said.

For each of the three tests, Brian's crew had to dig out rock and bury dozens of steel containers about eight foot square. Each had instruments inside, with pipes protruding from the roof above ground level. The ones closest to the bombs were sandbagged and concreted to protect them from shockwaves. Shortly after every blast, Brian - wearing only shorts, socks, boots and a hat - drove back in, removed the sandbags, busted up the concrete, and extracted the measuring devices. Scientists who went with them wore radiation suits and dose badges, but Brian had neither.

After Taranaki, someone decided to change that. 'After the third bomb, we were given little rubber boots, and a white overall, and a dose badge. We were told to walk through the crater. The mushroom cloud was still overhead. The wind had started to push it away. It was only a few hours after, not very long,' said Brian. 'As you approached the bomb site it was quite amazing because it was like a bowling green. Everything was green and smooth. It was only when you were on it you realised the heat from the bomb had crystallised the earth underneath it. It was a crust of molten sand, like glass. The crater left there was huge. They told us to walk into that, down into the crater, and up the other side, and then check our meters to see how high the dose was.'

He added: 'When our meters reached a certain point, they told us to come out. It didn't take long for it to reach that point. We weren't told at the time what the dose was supposed to be. But it was just as bad as going through the centre of the bomb as soon as it had gone off.' Brian was taken to a decontamination area, his clothes stripped off, and told to shower. 'We spent five or six minutes scrubbing away, then put ourselves in this meter. It was like standing on a weighing machine, and you push your hands through these bars to be tested. If a bell rang, you were still radioactive and had to go back in and scrub your nails, everywhere, in your hair. I had to do it three times. They didn't give us any more information.'

Six years later, Brian was medically discharged with a duodenal ulcer. Radiation can cause problems with the lining of the gut, and test veterans are 20 per cent more likely than other servicemen to die from stomach cancer. Brian said: 'It wasn't until later on that I started having skin problems. It would cover me from head to toes, rashes on my back, chest, legs, thighs. They used to come out on the palms of my hands. I'd get a little itchy blister in the centre of my palm, it would break and then spread over the fingers. I used to wear white cotton gloves to ease the pain and itching. The skin would go hard, then crack and bleed, and it would start all over again. I had that for 20 years. No doctor could work out what it was.'

Modern cancer patients are warned that radiotherapy using beta radiation can lead to radiodermatitis, which causes rashes, skin peeling, and ulceration. It is caused by the decay of isotopes, including plutonium and cobalt-60. Brian said: 'I would have a constant itch, all over, and had to take cold showers just to stop the itching and have something of a normal life. I got depressed, to the point where I didn't want to go and see the doctors because they just gave me the same old medication and it never did me any good. Then one day, after 20 years, it just stopped, as suddenly as it came.' Around the same time, Brian had an operation to cure his ulcer. It involved cutting the vagus nerve, which controls digestion as well as carrying sensory information between the brain and the skin's surface. 'I told all my consultants what was done to me out there in Maralinga, and asked if it was due to fallout. They all denied it,' said Brian. 'Nobody's ever done anything for us nuclear test veterans except withhold information from us. That place is still radioactive, it's in the soil for a hell of a long time, so what chance does a human being have?' Brian died in 2024, still without answers as to what had caused decades of agony.

Publicly, the tests had passed off without a hitch. The safety committee produced a report declaring 'stringent safety conditions were imposed', that 'detailed tests were carried out on sheep to confirm that no untoward biological effects had occurred'. They said the fallout was the radioactive equivalent of wearing a wristwatch with a luminous dial for a month; they did not mention weak fallout might still be capable of being ingested. Nor that their means of measuring it was basic. They had 'mobile monitoring stations' - jeeps driving around the range, covered with sticky paper - or 'stationary monitoring stations', which involved selecting a single roof in a distant city, sealing it, then collecting rain water from it. There is no evidence the committee checked water sources, or people.

They concluded: 'The Safety Committee has satisfaction in reporting to you that, after analysis of all these data, it can assert that the safety measures taken were completely successful. The fallout on the occasion of this trial was considerably less than in any previous trials and there was absolutely no risk to any individual, livestock or property at any time.'[60]

By this point, 12 nuclear weapons had been detonated on Australian soil over five years. It was quite an assertion to make, and entirely false. Blasts had repeatedly taken place in breach of the rules, fallout was found where it shouldn't have been, indigenous people were in the danger zone, and an ever-decreasing number of personnel were being monitored for radiation.

Before, during and after Antler, the Minor Trials continued. Much was made to the Australians that, as these were not nuclear devices, they were little more than conventional explosions. Discussing them in Parliament decades later, then-Prime Minister Margaret Thatcher said: 'A number of experiments with radioactive materials but not producing nuclear yields were conducted... In some of these, beryllium and/or natural uranium was used with a resultant minor contamination of the local environment; I am advised that the subsequent clean-up operations ensured that there was no residual hazard.'[61]

But in a nuclear blast, some of the radioactive material can be turned to pure energy, and if it's exploded high enough fallout can be limited. If anything, the Minor Trials were more dangerous - not only were they at ground level, but without a nuclear reaction to consume it, all the radioactive material fell back to earth.

Britain now knew exactly how to make a star – and she quite literally threw her best men at it. In 1958 Eric Denson was 26 years old, a flight lieutenant in

Above left: Eric Denson was a flight lieutenant when he flew a Canberra bomber through the very heart of a mushroom cloud in an 'experiment' conducted without his knowledge. (Courtesy of Shirley Denson)

Above right: Eric pictured with wife Shirley Denson and their first child. Shirley later became a strident campaigner for families of those who served at the nuclear tests. (Courtesy of Shirley Denson)

the Royal Air Force, and 'handsome and sexy as hell'. He was witty, charming, extremely popular with comrades, captain of boxing at the training college, and an exemplar of the 'excellent leadership qualities' that had seen him recruited to one of the toughest jobs in the forces. A group captain later recalled how 'proud' he had been to introduce this 'distinct credit to the system' to his sister Shirley, who was two years younger, at an end of term ball.

The couple fell in love and got married, and very quickly had their first daughter, Suzanne. 'He was one of the best pilots they had,' said Shirley. 'That's why they chose him. If you wanted someone to fly through a nuclear bomb, he was the one who would be able to do it.' Eric's job usually involved high-altitude patrols of the border with Iron Curtain countries in Eastern Europe. One day he came home and told Shirley he had to go away on a top secret mission for six months. Suzanne was just two years old, and Shirley was pregnant with their second.

Part of the problem with the failed Grapple tests was the purse-strings were never loosened. From pre-Hurricane days when the money was smuggled sideways out of other budgets, to the no-holds-barred race by world powers amid growing opposition, there was a reluctance to be seen spending too much. According to Oulton, 'in pursuit of the Ministry of Supply's firm directive to exercise the greatest economy, the whole operation was, in relation to the enormous task to be performed, run on a shoestring, and there was not much to spare of anything'. The British government had the chance to lease or purchase the island, which would have given it full authority but also full responsibility. It decided the offer was 'premature'.

One evening after dinner Cook and Oulton went for a walk along the beach, gin and tonics in hand. 'We haven't got it quite right,' Cook told the commander. 'We shall have to do it all again, providing we can do so before the ban comes into force. So that means as soon as possible; say two to three months.' Oulton pointed out that the island had only enough men and equipment to get through the tests they'd just completed. The 3,515 troops were all due leave. Morale, and obedience, would disintegrate if they were ordered to hang about any longer. And it was out of the question that HMS *Warrior* could stay as control ship, with an accompanying armada, for months.

'But there is an alternative,' Oulton recalled saying. 'From here to the south east tip of the island is nearly 30 miles. I reckon we could do the live drops at 8,000 ft, one and a half miles beyond the point, with only very minor blast damage to the camp and the airfield. What do you say to that?' Cook agreed, and they presented the plan to use Christmas Island - inhabited by more than 200 people who'd still be there when the British left, and temporary home to thousands of servicemen - to the Treasury official in charge of the money. It was riskier, because of the dangers of fallout if a plane crashed carrying a nuclear payload. If the RAF missed their target by even a small margin, the consequences could be catastrophic. But it was cheaper in terms of manpower and logistics. The Treasury approved it, so 16 years into the weapons programme Penney and his team prepared to bomb somewhere they knew somebody called home.

Brian Unthank signed up to the RAF as a trainee cook in 1956, when he was 17. A year later, the lad from Dartford in Kent was one of thousands put on the 'preliminary warning roll' for immediate deployment to Christmas Island. He was part of the reinforcements, and had a full medical before he left. It noted his urine, blood pressure, reflexes, lungs, heart, hearing, and eyes were in tip-top shape, and even that his gums were healthy.

'We flew in a Super Constellation over the North Pole to Canada to refuel, then on to Honolulu where we stayed in a hotel on Waikiki beach,' he remembered. 'I was 19, it seemed like a holiday. We got 200 free cigarettes a week and all the Tennent's lager we wanted. No-one mentioned nuclear bombs. We had no idea what we were doing or where we were going.' The bombs were to be dropped just 23 miles from the main camp, which was on the north coast of the island. Some building work had to be done to bring an airfield on the southernmost tip up to scratch, plus blast-proof bunkers for scientists and instruments. There were tented camps for the troops who maintained it all - placing them, in some cases, just a handful of miles from the drop sites. In addition to the dozens of islanders already being used as labourers, another 40 were brought in 'to meet the task force's needs'.

But Oulton had to fight for it. In August, documents show him badgering MoD officials for more kit, saying 'recent cuts... made it increasingly difficult to provide the support required for a megaton trial. Many of the installations put up temporarily... needed repair or renewal, for example, the tents would barely last through the November operation'. He asked for allowances to replace uniforms which had been ruined by the humid atmosphere and high rust content in piped water to the laundry, but was told that each serviceman would need to make an individual claim to the Treasury. Those who'd already returned home, with half-rotten uniforms, were told they could replace them at their own expense.[62]

Scientists, however, were treated with kid gloves. In February 1958 the finance committee agreed to allow an extra sixpence per person, per day, to improve the rations. It came after AWRE staff had refused to return to Christmas Island because the food was so bad. It added £30,000, or £739,000 in today's money, to the catering bill.

They were just as tight-fisted with equipment. The order went out for two Canberra PR9 bombers to conduct cloud sampling at a cost of £250,000 each, or about £4.9m today. When it turned out they wouldn't be ready in time, two older PR7s had to be taken out of storage, at half the price. The Valiants were also upgraded with £50,000 of navigational equipment, to ensure pinpoint accuracy.

The lack of a naval fleet meant limited radar surveillance. Oulton told the Treasury that without warships, he'd need a £200,000 Marconi radar system. He only got it after he said Grapple X would have to take place whether it was there or not. HMS *Messina* was so old that, despite £30,000 of repairs, within months she would no longer be seaworthy. So another £200,000 was spent refitting HMS

Brian Unthank, pictured with a poppy wreath to lay on behalf of campaign group LABRATS, at Horseguards Parade on Remembrance Sunday, 2023. (Susie Boniface)

Narvik as well. And at AWRE, the boffins were urged to find savings too - in November 1957, they said they might be able to bring the price of plutonium down to £25,000 a kilo.

Meanwhile, Brian and his colleagues fed and watered the troops every day, sometimes with fish they'd caught. He recalled being camped about 20 miles from Ground Zero, close to the beach. 'We hung a hawser cable from a palm tree outside our tent, hooked some meat on it, and tossed it over the reef,' he said. 'We caught sharks, and if they were big enough we served it in the mess. If not we'd eat them ourselves. It was delicious - tasted like beef.'

But there remained the problem with morale. Oulton noted: 'The cheerful put-up-with-the-snags-and-get-on-with-this-important-job attitude of all ranks was changing to a sullen resentment. The troops of all three services had had a pretty miserable time, despite all efforts to the contrary, but had been buoyed up by the belief that the task was of great national importance and the sooner they got the three tests done, the sooner they could go home.' Some of the men were given shore leave in Fiji or Hawaii; officers organised movies, projected onto a painted wall in the main camp, and sports events. There were periodic briefings about the importance of their work and absolute safety of it, but the bursts of patriotism were punctured by the arrival of 'Dear John' letters from girlfriends at home. They had heard the boys were to stay away longer, and in Britain there was public debate about radiation's likely impact on future generations in the event of a nuclear war. Some girls decided they would rather settle down with someone else.

A planning document found later in the Public Record Office recorded preparations for the blast. The local islanders were to be boarded on one of the warships anchored off the main camp, which would be ready to sail. 'Should an accident occur, either due to a crash on take-off by the bomber, or a surface burst instead of a high-air burst, then there may be a risk to ships lying in the anchorage,' it said. 'This risk is not immediate and there will be ample time to direct the ships to move to avoid the risk. The advice given by those arranging the operation is that these arrangements will positively assure the safety of the Gilbertese on the island.'

If true, the ease with which any vessel could merely sail away from the nuclear weapon that had already exploded might have made the bomb a little less terrifying to all concerned. In any event, the report insisted the public mustn't know what was happening. 'In view of the secrecy attaching to the operation, we would obviously wish to delay an announcement till the last practicable moment,' it said.

Word still seeped out. In October 1957, the Japanese ambassador wrote to the Foreign Secretary, formally requesting that if any Japanese suffered damage or loss as a result of the nuclear tests, they should be able to claim compensation from the British. The response was curt: 'Any person who knowingly entered or remained within the "danger area" would, in the opinion of Her Majesty's Government, be unable to establish any claim to compensation for any damage or loss which he might incur.' The writer did not bother to pretend, as the government did to its own troops and politicians, that British servicemen ordered into that 'danger area' would not be at any risk. The Japanese already knew better than anyone else what nuclear bombs could do.[63]

Grapple X was planned for the beginning of November 1957, but they were now entering the stormy season. The drop was delayed for three days by bad weather. With each postponement, Shackletons had to restart patrols of the testing area, soaking up fuel and flying hours. To overcome it, everyone was put on 'full readiness and standby through all the daylight hours', so they could do a test at the first break in the weather. On November 7, it was decided the clouds should clear the following morning, and some of the islanders were flown to nearby Canton Island. Oulton said they went 'not very happily this time but they just had to be out of any harm's way, with this first live test over Christmas itself'. Troops, however, remained.

This time, there were no media to see any embarrassing failures, and only two US observers. At 1am, Oulton was told one of the patrol planes had spotted a cargo ship, the SS *Effie*, nonchalantly sail into the exclusion zone. The British government had delayed announcing the no-go areas, and the vessel had left port before the official warning was issued. The Shackleton's wireless operator was trying to contact her without success, and Oulton ordered *Cossack* to intercept with all haste. As dawn broke around 6am the Valiant crew was ready to take off, but there was no word from the *Effie*.

At 7.20am the Valiants were ordered to start their engines. Ginger Weir, for whom as senior RAF officer a lot was riding on this, was still worried about the weather and a working explosion that could also be successfully checked by the sniff planes. 'Aren't you pushing it a bit?' he asked Oulton, who replied: 'If we stop now, it'll take us a month to get ready again. I reckon it's just about all right and we'll press on to the last moment before calling it off.' The bombers took off at 7.35am, and were in the air when word came through: the crew of the *Effie* had woken up and turned on the radio, and finally heard the frantic messages about getting the hell out of there. They left under escort from *Cossack*, and the drop could go ahead.

Grapple X was a rethought version of the earlier attempts, this time with a mix of plutonium and uranium in the first stage of the device, supercharged explosives to split the atoms, and with three layers in the outer jacket to increase the compression and create fusion. The bomb was dropped at 8.47am, and this time they'd done it: 1.8 megatons, a true thermonuclear weapon at last. As soon as the flash had passed, Oulton rushed outside to 'get some firsthand experience of the blast wave'. He described being 'bowled over like scraps of straw in a gale'. He said: 'It was worse than being at the bottom of a Welsh rugby scrum in a needle game.' The blast smashed through hatches and windows of helicopters parked up on the airfield.

Brian and his mates were ordered to crouch on their haunches under palm trees, about 75 yards from the mess hall in main camp. 'I was wearing a shirt, shorts, socks and boots. Maybe a floppy hat,' he said. 'We had to kneel down with our knuckles in our eye sockets. They counted down, and I saw the white flash. If you live to be a million years old you'll never ever fully describe its brilliance, the depth of it, the intensity. I could see every bone and blood vessel in my hands.' Told to stand and turn to look, Brian saw 'a ball of fire, black, white, red, and a big black line, hurtling towards us'. It was the pressure wave, and it threw him to the ground.

Gordon Coggon was there at the same time, as RAF ground crew. When the bomb went off, he was sitting cross-legged with his comrades, his beret held in his hands and pressed into his eyes. 'I wanted to run, men were shouting, some crying, it was terrifying,' he said, before the tannoy announced they could turn and look.

'You could actually hear men sighing with relief.' Later he was ordered to wash down one of the Canberras, wearing denims, wellies, and a pair of rubber gloves. He was given a full face mask with a breathing tube, but the straps were faulty and he was ordered to switch it for a mask that just covered his nose and mouth. 'It was useless because when it got wet from the splashback of water, I had to pull it down off my face in order to breathe,' he said. 'Afterwards I went into decontamination, which took almost 1.5 hours. I was told to put the clothes I had worn into a yellow 45 gallon drum which had the radioactive symbol on it. I went sick later because I felt washed out and a week after I had to have a series of boils lanced on my back and neck which continued to come back over a period of three months or more. I very often wake up from nightmares where I am lost in white fog caused by the nuclear bomb and feel like I am suffocating.' Six years later he developed scoliosis, gut problems, needed to have all his teeth removed, and his gall bladder out. He later developed prostate cancer and diabetes.

With the H-bomb achieved, the pressure eased a little. There was even an upside: on December 4, 1957, Oulton reported to the Atomic Weapons Trials Executive that moving the test to Christmas Island 'had made the mounting of the operation much easier and had resulted in savings which might be as much as £2m'.[64]

This photograph was produced in Parliament by Tamworth MP Brian Jenkins in a debate in 2002. It shows his constituent Ron Coates with three comrades aboard HMS *Warrior* in May 1957. Two of the men developed cancer, and a third diabetes. Ron was told that to qualify for a war pension he would need to prove he would not have had cancer if it was not for the radiation. Hansard, December 4, 2002.

An official photograph of the Grapple X cloud, Britain's first thermonuclear bomb. (Wikimedia Commons)

But he vented about how close they'd come to disaster with the *Effie*, in a separate cable to Whitehall. 'After the most strenuous effort to overcome heartbreaking difficulties, we had a fleeting chance to fire this morning. This opportunity was very nearly lost,' read his telegram. 'I wish to register the strongest possible protest against such gross interference with such a difficult and dangerous duty.' Yet he was about to ship in a group of people rarely seen in modern military operations - dozens of wives and children.

There was nothing the MoD could do about sweethearts going cold, but they decided to help the married troops. Whitehall bankrolled an idea from Colonel Henry Gatford, the commanding officer of 25th Field Regiment, Royal Engineers, whose 400 men had been on Christmas Island for a year and were some of the lucky few who were allowed home. The troopship *Dunera* was bringing out replacements, and at Col Gatford's suggestion the wives and children were given the option to come and collect their men, at a subsidised rate. For £30 an adult and £15 for a child, they were promised a 'ten-week tropical cruise that would cost anyone else hundreds of pounds'.

So as Christmas 1957 approached, 35 children, and 31 wives were shipped out as a morale boost, alongside a company of replacement Royal Engineers. Among them were Sadie Midford and her daughter Valerie, one of five babies on board. Sadie recalled 50 years later: 'Her dad had left soon after she was born, so I leapt at the chance. We had a great time on the ship and joked about how when we got home our men would be glowing... but we had no real idea.' The ship arrived in January 1958, 10 weeks after Grapple X was dropped. They spent several days at Christmas Island, the children playing in the surf as the vessel unloaded and repacked for the trip home. On the way back, the now-reunited families stopped at Hawaii, where *Mirror* photographer Alma Sturney took photos of the children, wives and troops building sandcastles, dancing with islanders, and wearing flowers in their hair and around their necks. The paper dubbed the *Dunera* 'the happiest-ever troopship'. Sturney wrote: 'Husbands and wives, who had last gone shopping together over a year ago in grey city streets, set out under the sun to buy gay Hawaiian-print cottons and dazzling sports shirts... the regiment's single men got chatting with the hula girls... when the *Dunera* sailed, the decks were lined with British mothers, fathers and children blowing kisses landwards.'[65]

On the way home, Sadie noticed Valerie's hair was falling out. When she later told their doctor where they'd been, he was aghast and refused to believe her. Aged 31, Valerie had part of her cervix removed because of pre-cancerous cells. Sadie said that before they boarded at Southampton, she was told to attend a local hospital for a medical. Blood samples were taken from her and both children. It is not known who holds the records.

Liz Burridge was 13 months old when she was taken aboard the *Dunera* to collect her dad Peter, an army engineer. 'It probably seemed like a good idea at the time, but looking back now we were just propaganda for the MoD. They took our pictures and we were in the *Mirror*, on this lovely holiday to the tropics. We went ashore and played on the beach,' she said. Letters written by her mother Joan show that the toddler had inexplicable and dramatic weight loss on the return trip. Shortly afterwards, Joan fell pregnant. But she gave birth to a severely-deformed boy, who didn't survive. Liz said: 'I only heard her speak about it once, but she was extremely upset. It wasn't something you discussed in those days. But they never

Sadie Midford disembarks the *Dunera* in Honolulu, carrying baby Valerie, with other wives on the way to Christmas Island. (Courtesy of Valerie Chir)

tried again, and years later my dad told me "you should have had a brother".' Joan died from a brain aneurysm aged 54, and Peter from prostate cancer in 2016. Liz herself was unable to have children, and was later diagnosed with an aneurysm herself. 'My father was always annoyed at the lack of recognition for the veterans. My father always said the bombs were a deterrent, to make people stop and think. Well, they need to,' said Liz.

Efforts to keep the troops happy weren't all successful. In February 1958, a month after the *Dunera* had arrived home and as the freshened-up task force was preparing for another H-bomb, the House of Commons held an urgent debate to discuss complaints about living conditions on Christmas Island. Labour MP Richard Winterbottom read out letters home from the troops, which had been sent on to him by their worried parents. He read out a poem from one servicemen, which said: 'The isle it abounds with monstrous ants, which infect our clothing, our shirts and our pants. And since we came here we have done nothing but curse, for even Alcatraz could not be worse. For even convicts do get their rest, but for us it's a ruddy endurance test.'

He said the MoD had known conditions were poor for months, that troops dug a hole to use as a toilet rather than enter the foul latrines, and at one point an officer was pelted with rotten food in the RAF mess. 'According to almost every letter

I have received, the men find rats, ants or crabs in their blankets at night,' said Mr Winterbottom. 'The rainfall and the winds, according to many letters, are so bad that on two nights out of every three the tents are blown away, and on the third night they are waterlogged... The only refrigerator is on the ship, which brings the food either from this country or from Singapore. Every letter I have received says that it is rotten, and some add suitable adjectives in support... Prices in the canteen are prohibitive, which offsets any extra pay they may get. For a quarter of a pound of sweets for which people are charged 8 pence in this country, the soldiers have to pay two shillings and six pence.'[66]

Junior defence minister Charles Orr-Ewing blamed the discontent on a 'lurid' letter home from one corporal which was published in the newspapers, and had distressed wives and families. He said new tents had been sent out, and each would be on an individual base to raise it above ground level to keep out the wildlife. 'The most serious trouble has come from men cutting themselves, when swimming normally or taking part in underwater swimming, upon the sharp and jagged coral which exists round the island,' he said. 'Life might get boring on the island, and it is, therefore, right that plenty of amenities should be provided. There are a dozen football and hockey pitches, three or four cricket pitches, and a sailing club, which is used almost entirely by the men. Until recently only one officer was a member. All the men can indulge in swimming, both ordinary and underwater, and there is also first-class fishing.' When the Labour MPs suggested he visit the island to see the conditions for himself, the minister said he'd be happy to, but Parliament couldn't spare him.

Billy Morris was the first to die – a healthy lad of 20 from South Wales, reduced to skin and bone only six months after being rated 'A1 fit'. His sister Rachel said: 'I remember him in his uniform – he was so handsome and such a lovely boy. He was never sick a day of his life. He went to Christmas Island on National Service and a few months after he came back, he wasted away to nothing and died.'

Billy was a sapper in the Royal Engineers. He was seen by five doctors when he enlisted, who recorded he was in peak physical condition. Rachel said they also took blood tests, which would be standard procedure. He was sent out after the first Grapple failures, and was one of 2,300 to see Grapple X in November 1957. He reported sick to the unit medic on January 1, 1958, with nosebleeds and swellings in his neck.

On January 20, he was flown to a military hospital in London, where he was diagnosed with 'a blood disease' and at the end of February was transferred to a hospital in his home town of Swansea, where they decided 'he was suffering from an organic disease of the glands, and after treatment he was sent home'. He was also discharged from the army on medical grounds. But Billy collapsed at home on April 30 and was readmitted. He was diagnosed with leukaemia, and died on June 14. His girlfriend Sylvia was pregnant. The *Western Mail* reported two days later that the post mortem 'states that Morris died of leukaemia not connected with his sojourn in Christmas Island'. How any pathologist could reach such a definite

Sapper Billy Morris was a strapping young lad who was rated A1 fit, but died from leukaemia just months after the first H-bomb. (Courtesy of Rachel Morris)

conclusion as to the cause of a blood cancer, without benefit of modern techniques, is not clear.

A bone from Morris' body was sent to the radiobiological research unit at the AWRE, which found he had a 'normal' amount of strontium-90 in his skeleton.

But at the inquest Doctor Owen Glyn Williams, asked if Morris' death was caused in any way by radiation, said: 'What one could say is this: that this problem is under review, but there is no doubt that the incidence of leukaemia is increasing, and that it does bear relationship to this fallout throughout the world.' Questioned further, he added he knew of no case where leukaemia had developed within six months of exposure. He then went a step further, and turned a lack of knowledge into definitive proof: 'In the light of what I know, I would say there is no connection with this boy's disease and the fact that he was involved in the atomic explosion on Christmas Island.'

Today, it's known that there are many different types of leukaemia, some of them directly linked to radiation, and caused by changes in DNA. In the wake of Hiroshima and Nagasaki, doctors first noticed an increase in blood cancers two to three years after the blasts, and they tailed off after about 25 years. Without genetic analysis to determine exactly which parts of his DNA had gone wrong, it was not possible to be definite about the cause of Billy's cancer. So the coroner and pathologist relied, instead, on the circumstantial evidence of strontium in his bones, treating it as an indicator of any and all radiation Billy may have absorbed. While it may have been 'normal' for the time, it does not mean his cancer was unconnected, nor does it rule out a different radioactive cause for his disease. In 1958 they simply couldn't be sure - but that didn't stop them saying they were.

The *Western Mail* reported that Billy was attached to 'the atom bomb disposal unit' and 'he was wearing protective clothing'. But his inquest heard Billy's protection at the tests was limited to a pair of dark glasses. His commanding officer, Major Norton Barnes, told the inquest that Billy never handled radioactive material, and 'did not at any time have to wear protective clothing'. He added that radioactivity in the areas where Billy worked was 'nil', but gave no evidence to support it. He confirmed that rainwater entered the men's tents after the blasts, but said 'the beds were all at least 12 to 18 inches above the ground'. Major Barnes added: 'The distance we were from the explosion had been declared to be safe by scientists and by the task force commander.' The exact distance was withheld from the inquest jury 'on the grounds of national security'.[67] An internal MoD memo notes the coroner had requested evidence from the AWRE, but it was 'not prepared to provide a witness'. Instead they worked up a PR line for the minister to take if there was any Parliamentary interest.

One medical specialist, who had served at Hiroshima, told the inquest that Billy had 'the wrong type of leukaemia' to be linked to radiation. But no explanation was given about how, if he had the condition before going to Christmas Island, he had passed blood tests and been rated 'A1 fit'. The verdict was 'death from natural causes', but studies have since shown there may be a connection between radiation and the type of leukaemia Billy was diagnosed with. 'The Royal British Legion wanted to use Billy as a test case for compensation. But his girlfriend lost the baby and my mother found it all too upsetting, so we decided to pull out,' said Rachel. 'We don't care about compensation - the most important thing is that the Government admits what it did to our Billy.' Rachel kept an eye on the campaign for recognition over the years, yearning for her brother's death to be acknowledged. In one of her letters to the *Mirror*, she wrote: 'When I am gone, who is going to remember the boys on Christmas Island who gave their lives to the army just as much as any other soldier today? These boys have no medals, or recognition, no parades to remember their passing. They lived in appalling conditions on a soaking wet, rat-infested island... all they get from the MoD is denial.' It was signed 'Rachel Morris, Billy's sister'. She died in 2022, aged 81, still waiting.

Other servicemen also died after Grapple X. A Lieutenant DC Franklin, who served on HMS *Warrior*, succumbed to aplastic anaemia within a year of the blast. It's a known radiogenic condition in which the bone marrow is unable to produce enough blood cells for the body to work normally. The *British Medical Journal* of September 1958 records that his inquest had been opened and adjourned 'to allow examination of his bones for abnormal radioactivity'. Derek Fiddaman was also there in 1957 – a 21-year-old naval rating on guard ship HMS *Cossack*, stationed west of the island. He was ordered on deck to watch the explosions, and believed prevailing winds later dropped fallout on the crew.

He said: 'I plotted the position of the ship at the point the bomb was dropped – and the wind was from the east.' In 1975 he developed lumps on his face that were diagnosed as basal cell carcinomas – cancers formed in the top layer of skin. In the decades that followed he had 1,000 hospital visits and more than 200 tumours removed. Derek said: 'I have about five cut out every year, so if I live another 20 years there'll be 100 more. Each time they cut one out I go back and there's another one growing in the scar.' After a long fight with the MoD he won a 60% war pension on the grounds his cancers were caused by radiation – from the sun, 93 million miles away. He said: 'I'm one of the lucky ones. It's not going to kill me. All I want is to hear the word "sorry". They never will, but that's all I want.' Derek died, aged 86, in 2022.

As the tests progressed, there were fewer safety checks. There were fewer dose badges handed out, and veterans who had overalls and balaclavas at earlier tests now wore everyday shirts and shorts as they were mustered to witness the bombs. Perhaps it was because most dose badges from earlier tests had been returned showing a nil dose - and perhaps it was because of the cost-cutting. Journalist and author Nic Maclellan, who has reported on the impact of the tests from the perspective of Pacific islanders, found in the Colonial Archives a host of correspondence between local officials and plantation owners. 'They were worried about the possibility of fallout affecting their crops,' said Nic. 'So the maps were redrawn. Before the letters were sent, the possible area of fallout is a circle around Christmas and Malden islands. After the plantation owners kicked up a fuss, the danger area was redrawn - in a square. And the plantations were outside it.'

As 1958 dawned, the Cold War was reaching a peak. The Soviets had beaten the Americans to space with *Sputnik*, a low-orbit satellite, and then sent up a Moscow stray called Laika in *Sputnik 2*. The US had just detonated 29 nuclear weapons in Nevada in Operation Plumbbob, and was planning more. The USSR now suggested a moratorium on testing, and Britain and the US sent experts to debate it while rushing to finish their explosions.

Grapple X needed a lot of enriched uranium, which was expensive to produce. Grapple Y aimed at using less of it by increasing other radioactive components. This new bomb was dropped at 10.05am on April 28, 1958. Brian Unthank remembers being in the NAAFI, and the blast knocking cans off shelves, and the Queen's portrait from the wall.

Penney had a military personal assistant, known as a batman, called Ralph Gray. He too was mustered on the beach to watch the largest weapon Britain had ever exploded. He said: 'I used to hear the officers and boffins talking at the dinner table while I served them for the four weeks they were there. I can't remember the precise words, but I was left with the impression they didn't have a clue what would happen, and they thought it might go wrong.' Ralph had two healthy children before he went to Christmas Island, but after his return had a third child with spina bifida, and a grandson with a cranial deformity.[68]

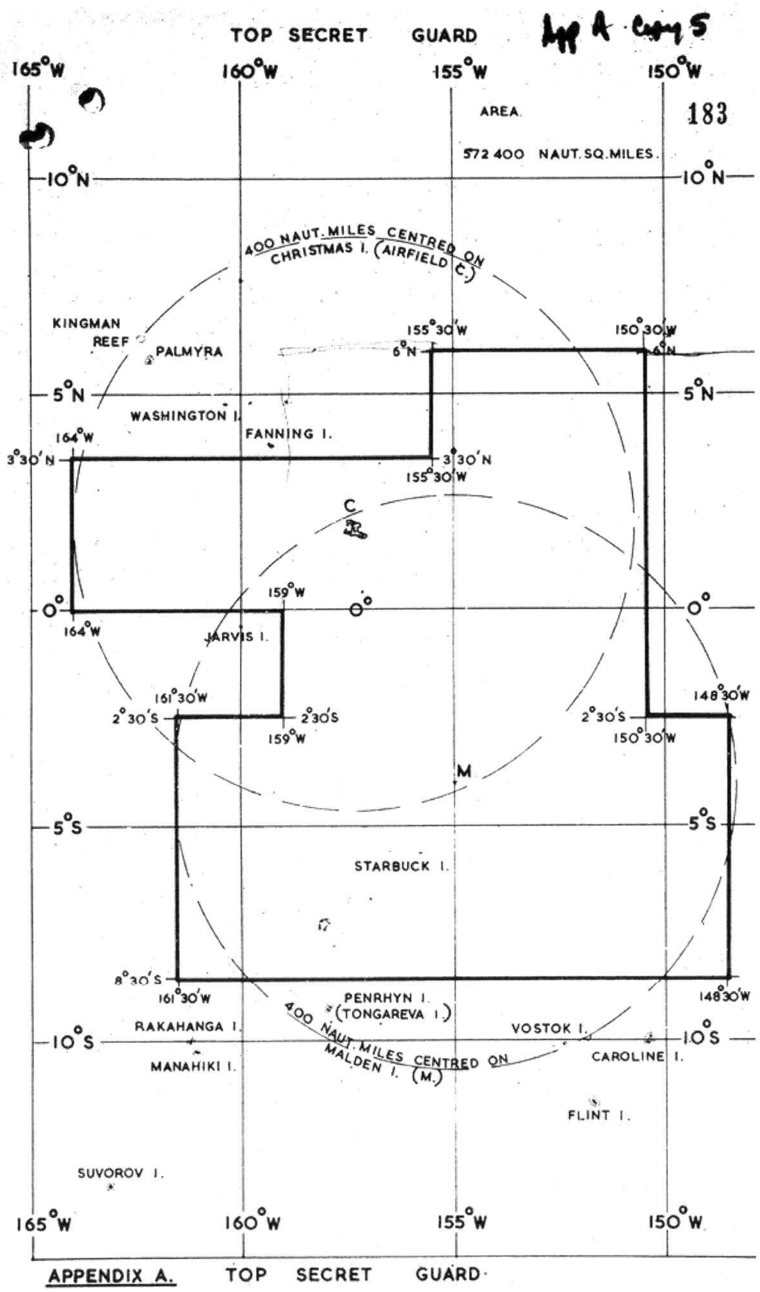

The official map, produced by the British for planning purposes, showing that fallout was predicted to fall in neat lines in order to avoid inhabited islands like Jarvis, Fanning, and Penrhyn. (Courtesy of Nic Maclellan)

An official photograph of the Grapple Y mushroom cloud, the most powerful weapon Britain has ever fired. (Wikimedia Commons)

Flt Lt Denson's Canberra was one of the 'sniff planes'. Records show he entered the mushroom cloud 49 minutes after the blast, and spent quarter of an hour flying through the incredible turbulence. Shirley remembered: 'Eric never discussed with me what had happened on Christmas Island. I once overheard him talking about it with my father. He said he had almost lost control as the plane went into the mushroom cloud. They were tossed about by the most incredible forces. He said he just held on, and recited that line from the Charge of the Light Brigade, "Into the jaws of death, into the mouth of hell, rode the six hundred…" over and over again.'

This time the bomb went according to plan. It had a yield of three megatons, 150 times greater than Nagasaki. A second time proved it was no fluke, and Britain was in the thermonuclear club at last. Later that year, AWRE scientists wrote a document titled: 'Determination of the whole body dose of Radiation acquired by Aircrew members while on Cloud Sampling detail.' It compared readings in the cockpit 'Charlie meter' with film badges worn by the three crew. It says: 'Care was taken to ensure that as little shielding effect as possible was given by the ejector seat and that no equipment of any description shielded the badges.'

Four badges were placed around each crew member – one on the headrest, one on each armrest, and the last 'in seat pan 1 inch from testicles'. It says using feedback from previous missions, the second 'sniff' plane was able to stay in the mushroom cloud longest, and states: 'Thus the initial experiment was carried out on personnel flying the Secondary sampler.' This was Eric's plane. The experiment proved the Charlie meter was inaccurate, and found the pilot was the most exposed crew member. The badge behind Eric's head recorded a dose of 18 Roentgen, roughly equivalent to 1,800 chest x-rays. The badge next to his testicles had a dose of 8.8 Roentgen, or about 880 chest x-rays. The scientists were aware Eric's family might be affected. Their report says the lower testicular readings 'are of interest and cheer in relation to the possibility of genetic damage'.

After the flight, Eric vomited for 48 hours. His log book shows that he went back into the same plane two days later on April 30 for 'radar calibration and formation flying'. A navigator on his squadron, Joe Pasquini, claimed the plane had not been decontaminated but left alone on an isolated patch of airfield. He also said using an expensive, and radioactive, bomber to calibrate ground radar would be ridiculous when there were many smaller, safer aircraft around.

He added: 'The only reason I can think of to send Eric back up, with another plane to observe him, would be to see how an RAF pilot and plane would operate after being irradiated. Seeing as he was used in an "initial experiment", perhaps that was the second.' The MoD says it has no record of the flight. Eric's logbook was signed off by his squadron leader, Terry Gledhill - who later implied to his family that Eric had taken the plane up without approval. Perhaps he signed the log to protect one of his men, thinking he'd had a mental wobble while under incredible strain.

Joe spent decades investigating what he believed was a cover-up. In 2019, he self-published a short book called *Flying Between Two Suns*, describing how on the day of each blast the crews awoke at 1am, breakfasted on eggs,

RAF navigator Joe Pasquini spent decades investigating the death of his comrade Eric Denson. He suffered more than a dozen cancers before dying in 2021, possibly unaware his research had uncovered evidence of experimentation on British troops. (Courtesy of the Pasquini family)

bacon, toast and tea at 2am, and had hours of briefings and pre-flight checks before starting engines at 9am. He was one of the few men who actually saw a thermonuclear detonation. The blinds were drawn down inside each plane, but in Joe's - codenamed 'Sniff Boss', and piloted by Gledhill - he as navigator had to peer through a slit of welder's glass to see what was going on. 'Using the one-eye technique developed during night flying operations - one eye was kept open, the other closed so if the bright flash burnt out the retina in the back of the eyeball, the other eye would hopefully still be available to see with - with one eye tightly closed and the other slightly opened, I timed the duration of the entire detonation process to a 19 second flash, from start to finish,' he wrote. 'This was the length of time it took for the hydrogen bomb detonation to expand from the size of a small motorcar, to a molten, swirling, perfect, round white ball of liquid nuclear magma, over one nautical mile in diameter.'

Joe wrote: 'A new Sun had been created in the atmosphere... a mile above the surface of the sea. I watched with awe and wondering at what was happening. It was like looking at, and then seeing, the face of God.' Now Sniff Boss 'raced towards' the cloud. They lifted the shutters, and began circling, waiting for the order to make their cut, said Joe. 'While I was looking down at the [explosion] below, I felt the heat of the real Sun on the back of my neck. It was eerie... then I realised, there were two Suns. One seven miles directly below me, the other 93 million miles directly above. I was flying between two Suns.'

It took them 6 minutes and 31 seconds to fly the length of the cloud, making it 47 miles in diameter. The stem itself was 8 miles wide, and still growing. As they approached the stem, the plane was hit by rainfall, something which should have been impossible at that altitude, where Joe had recorded the outside air temperature as -74 Celsius, so low that water couldn't be liquid. 'All the radiation instruments lit up like a Christmas tree,' said Joe. 'Twenty seconds later the rain and radiation all stopped simultaneously, just as fast as it had started.'

On the ground, veterans saw the same phenomenon - a sudden cloudburst from a sky with only one, massive cloud in it, and they reported the rain was black. Twenty minutes after the blast, at 40,000ft, Joe's plane entered the cloud. The radiation instruments were 'hard against the stops' for 10 seconds, then settled back. After 30 seconds they flew back out again, preparing for another cut of a minute's duration. Then the word went out to the circling planes, one at a time, to make their own runs. Sniff One took one side of the cloud for just over two minutes. Then the Air Controller - a boffin from AWRE who was in the Sniff Boss plane - ordered Sniff Two, with Eric at the controls, to fly through the dead centre of the cloud, at 51,000ft. Joe's memoir records that the Air Controller knew about the experiment being conducted on Eric's crew with the dose badges, and implied that was why he was sent into the very heart of the blast. 'The plane hit the very dirtiest and most radioactive part of the cloud, with radiation rates as high as 280 Roentgen per hour,' said Joe, who logged the readings that were broadcast by Eric's plane. The gasbag in the belly of the bomber was maxed out, while of the two meters inside the aircraft, one was giving readings 50 times what the other showed.

The squadron's operational record book states: 'Sniff Two obviously struck the hottest zone and slightly exceeded his maximum permitted dose in the course of only one penetration.' But Joe said that Eric's plane made three long passes through the cloud: one of just over 3 minutes, another lasting 5 minutes, and a third for six minutes. Fourteen minutes in total. Back on the ground, Eric's plane was many times more radioactive than the rest of the squadron. Scientists were delighted, noting 'the cloud samples... were much more active than anticipated'. Gledhill's plane was also over the limits, and the squadron log book records it caused problems by having to 'replace two crews instead of the one philosophically accepted as "expendable".'

Joe emigrated to the US and went on to suffer bouts of cancer almost a dozen times - of the lung, brain, skin, bladder, and prostate. He finally died in August 2021, aged 88, after many years of suffering and campaigning. A naturally cheerful man, he had regular bouts of understandable fear and 'prepping', with anxieties surfacing after the Three Mile Island reactor meltdown in 1979, and movies about nuclear apocalypse. He would plot the fallout zones to make sure his family were safe.

After Grapple Y, scientists found fallout at nine sites on Christmas Island. A Shackleton aircraft on the day of the blast found radiation 'on various sorties' at 300ft up, but the readings were hard to understand because there was so much contamination. 'On return to Christmas Island the aircraft was found to

Sqn Ldr Terry Gledhill took this photo of the developing Grapple Y mushroom cloud from his cockpit, as he flew in the stratosphere. (Courtesy of the Gledhill family)

be contaminated', said a subsequent report, 'the dose probably being received sometimes from contamination of the aircraft, from fallout on the sea, and from radioactive material still in the air.' The radiation counts were 'off-scale', and the plane was considered 'completely unsatisfactory' for the task of monitoring. Boffins asked for one with a pressurised cabin and recirculated air instead - with no mention in the report of what had befallen the Shackleton's crew, who had flown through heavy contamination without it.

A helicopter was sent out to collect a sample of sea water two hours after the detonation, with a report saying it was taken from 'the point of maximum water reading'. It was found to be no more radioactive than any seawater - but only because it was 'impracticable' to get a helicopter to the area of fallout quick enough to measure it, before it fell below the waves. Today, the MoD merely says that no radiation was found. The report added 'a very real fallout' was found in the south west of the island, nearest the drop site, 'confirmed by survey on foot'. The report could not say 'whether the contamination was the result of dry fallout or rainout in subsequent bad weather'.[69]

'You are about to experience something no-one else in the world will ever see,' the senior Army officer told 21-year-old sapper Raymond Webber and his mates. He not only witnessed five nuclear bombs, but one of his tasks was to drive a truck through the fallout to collect heavily-contaminated equipment while wearing nothing but shorts and sandals. Raymond said: 'A couple of senior types came and grabbed me one day, told me to drive them into the forward area. That was the sharp end of the island where all the bombs were exploded. It was a huge truck, kicked up tons of dust, and we had the windows down because it was so hot. We drove to a shed that had three generators in it to run the monitoring equipment that took readings during the explosions. I waited while they loaded some stuff in the back. I asked what it was and they said I didn't need to know. When we left there was a mobile decontamination point that they had to go through, and they had shower after shower because their readings kept showing radiation. They washed me down too but we all put our uniforms back on, which must have been contaminated. And the truck wasn't washed, and we all climbed back in that.'

On returning home he married and had children, leaving the Army in 1965. He said: 'It wasn't long after I came back that I was wearing false teeth, and I've got osteoarthritis in my knees and back. But I'm not worried about me. It's my children.' He had five sons, all now in late middle age. The oldest was diagnosed with arthritis while undergoing Army basic training, and now has crumbling discs and cluster migraines. The second son has arthritis, a deformed lung and chronic condition lymphoedema. Another has generalised arthritis and a thyroid condition, while the youngest child is the healthiest.

Third son Philip, 60, has arthritis in all major joints, stage three chronic kidney disease, diverticulosis, cellulitis, depression and insomnia. His daughter, now in her 30s, has pain in her knees and hands already. Raymond said: 'Radiation knows no time limits. None of us knew what we were letting ourselves in for, and we had

Raymond Webber, pictured soon after joining the Royal Engineers and before he was sent to Christmas Island. (Courtesy of Raymond Webber)

absolutely no protection. Ever since we've tried to get people to believe that what went on should never have happened.'

In Geneva, arguments about a testing moratorium were finishing. By August 1958 they had a plan for monitoring stations worldwide that would enable the US,

UK and USSR to keep an eye on each other, preventing any future atmospheric tests.

With a deal ready, the governments rushed to finish their explosions. The Americans and Soviets detonated 54 between them that autumn alone. At Christmas Island, there were still four more to come. Operation Grapple Z was aimed at refining the weapon to get as big a blast as possible out of a bomb weighing no more than a ton, before the ban in October. Pennant, which used different ingredients to prime the bomb, was suspended from barrage balloons. Flagpole was dropped from a plane using blind radar techniques. Halliard had a three-stage design to trigger the blast and used pilot visual targeting, while Burgee was exploded using barrage balloons and tritium gas. They had a combined yield of just over a megaton and were the last nuclear weapons Britain ever fired alone.

RAF sergeant Roy Kirkland was ordered to sleep just half a mile from the Grapple Z Ground Zero. In a memoir, he wrote about living in a tent next to a target area that was turned to glass by the bombs and said he was also told to clean and pack contaminated instruments used by scientists to monitor the blast. It reveals that after a few months his unit was ordered to move from the main camp to 'B site', right next to the target area. Roy wrote: 'Scientists were going to explode

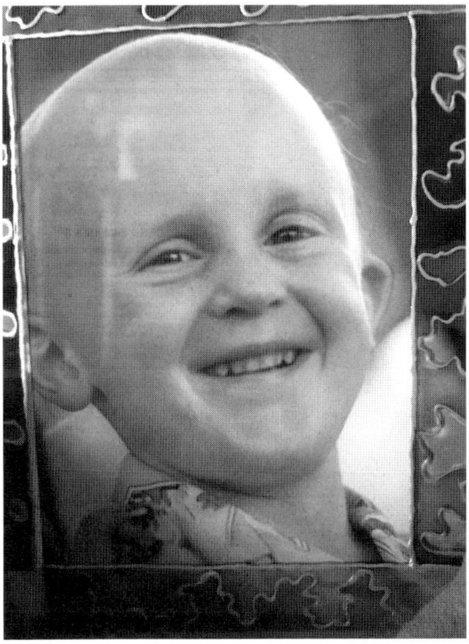

Above left: Wayne Kinson, whose grandfather Roy Kirkland had slept half a mile from Ground Zero in 1958, died from a cancer of the nervous system aged seven. (Courtesy of the Kirkland family)

Above right: Roy Kirkland pictured while on Christmas Island in 1958. (Courtesy of the Kirkland family)

an A-bomb. This was to be done by putting the bomb in a large basket slung under a flying barrage balloon tethered to the floor by steel cables rooted in large concrete pads. Now we knew what those concrete pads half a mile from our tent lines were for.' He said for the explosions they were in a level area close by the tents. 'We huddled together in total terror, stunned and silent by this phenomenon. There are not enough words to describe everything at that moment in time. I would say it took at least half an hour for us to contemplate even talking in any semblance of normality,' he said. 'What right had the British government of that day to subject me and thousands more to that torture without facts being explained to us first about the after-effects? We stayed at B site as life went on, writing letters home and yearning for the day when we could go home.'

Roy Kirkland pushing his grandsons, Wayne on the right, in a wheelbarrow at the family home. (Courtesy of the Kirkland family)

Roy returned home to Burton-on-Trent, his wife Angela and baby Denise, who was born while he was away. The couple went on to have four more children, two with chronic health problems. Daughter Melissa has severe thyroid and joint issues, and her son was born with a faulty heart valve. Another daughter, Claire, had skin cancer, chronic fatigue syndrome, and has unexplained lesions on her legs.

But it was Claire's son Wayne who paid the highest price. He was diagnosed with an extremely rare cancer of the nervous system, called a neuroblastoma, aged just three. He underwent four years of aggressive treatment before dying when he was seven. Wayne's aunt Denise said: 'My dad used to say all they had on was a pair of shorts and when the bombs went off they were simply told to look the other way. Afterwards he was told to clear the beach of all the dead birds and debris from the bombs using their bare hands. He was devastated to see the illnesses in our family and particularly when Wayne was diagnosed he would say "I wonder if it was anything to do with me".'

Squadron Leader Tony Davis was in a plane circling Christmas Island during the Grapple tests, taking photographs of the cloud for scientists. Six years later he was dead from acute myelomonocytic leukaemia, in which white blood cells multiply rapidly in the bone marrow. Immature cells, incapable of doing the job, take up perhaps a fifth of the room available for blood. Patients cannot fight off infection, become anaemic, and their blood stops clotting. It is known to be radiogenic. According to the NHS, 'being exposed to a significant level of radiation can increase your chances of developing AML, although this usually requires exposure at very high levels'. It can also be triggered by chromosomal damage, or having another blood disorder.

When Tony died, his sister Sue was his only living relative. Certain his death was linked to the tests, she enlisted the help of a solicitor to get the MoD to admit it, thereby absolving his estate of death duties. Had she succeeded, it would have meant many veterans' families would be better off, and all air crew, at the very least, would have been acknowledged as in danger and potentially compensated. Instead, in 1964 an MoD official wrote to her saying: 'I have been in touch with a medical branch on this case and they, after a very careful examination of all the records, have advised that during all trials conducted at Christmas Island, all aircrew were covered by some form of monitoring device to ensure that the safety procedures against accidental exposure were honoured. Without exception all such results were negative for all trials. Therefore it can be said that Sqn Ldr Davis was not exposed to radiation of nuclear weapon detonation origin during the course of his duties at Christmas Island... there are no other records of RAF personnel having suffered ill-effects from radiation.' Each of those sentences was later shown to be a lie. Sue died before she could prove it.

Five months after the last bomb, Christmas Island had a special visitor. Prince Philip, during a year-long Commonwealth tour on behalf of his wife the Queen, visited aboard the Royal Yacht *Britannia*. He was given a helicopter tour of the bombed areas, and officials made plans to land him at Ground Zero 'if requested'.

Troops were ordered to paint some of the rocks and sweep roads. The prince took his meals aboard the yacht, rather than in the mess.

When the prince arrived, Mike Rubery, a sapper in the Royal Engineers, was among those who formed the honour guard. In 2008 Mike, then 70, told the *Mirror* how after his time on Christmas Island he developed an ulcer and unexplained rash-like skin condition. He said: 'On Boxing Day 1958, for a treat, they took us for a picnic. We had a bag of sandwiches and a few cans of beer each, and they took us to Ground Zero. It was like a moonscape, all blackened and empty. We were picking things up and saying, "I wonder what this was?" - eating our food and drinking our beer. It sends a shiver down your spine to think of it now.'

Philip never asked the helicopter to drop him at ground zero. From the air, he would have seen the desert island's southern tip looking as smooth and verdant as a bowling green, with the vitrified, glass-like sand glinting in the tropical sun.

The final Grapple tests were not the end of the nuclear experiments, or the use of servicemen to study their effects. Kittens and Tims continued to run at Maralinga, refining triggers and cores. The incoming ban caused concern that it might mean an end to the Minor Trials. The British made an effort to frame them as comparatively harmless, with disingenuous memos between top brass discussing how to market them better. A. R. Bryant, the senior superintendent of weapons assembly at the AWRE, suggested the following definition: 'A minor trial is defined as a trial in which small amounts of radioactive or fissile material are involved in association with the detonation of conventional high explosive in such a manner that no fission results.'

In 1959, a new minor trial was launched called Operation Vixen, studying what might happen to a nuclear warhead in the event of an accident. These did not involve a nuclear reaction, but still dispersed the weapon and its contents. This material was collected at sampling stations, aircraft undertook 'sniff' missions, and detailed weather data was collected, often from balloons. Penney wrote to Titterton, concerned the use of plutonium would cause political problems. He explained it was vital to know about the dispersal of fallout during and after a fire, and the only way to find out was to start one. He asked how best to get Australian approval, and in response Titterton wrote to his government advising them to agree, so long as 'no material could escape' and full details were shared with him. When the formal request was made by the British government, it demanded no publicity of any sort. The Australian Minister of Defence, Athol Townley, agreed, saying 'I am not troubled very much by the trials in themselves' although it was 'potentially dangerous' and would 'bring the usual howl from the "ban the H-bomb" section of the community - Communist and otherwise'. To get around the disapproval, he suggested it be discussed between prime ministers, and kept secret from respective Cabinets. The trials went ahead in complete secrecy.

Vixen A ran from 1959 to 1961, with 31 experiments exposing nuclear devices to a petrol fire, inside an electric furnace, and using high explosives on them. Declassified documents later showed that scientists built a chimney, 11ft high and 4ft square, in the desert. Metal rods were put in the chimney for defined periods, totalling 70kg of uranium, 6kg of beryllium and just under a kilo of plutonium, at temperatures ranging from 600 to 1,000 degrees Celsius. The resulting fumes were channeled up into the air, to disperse where the winds took them. Other toxic materials were also used, including polonium-210 and actinium-227, a trace element from uranium decay. Monitoring stations were set up downwind, to measure how much got how far. The resulting fallout pattern produced from the readings shows it spread across a vast area, including well beyond the monitors.

Twelve Vixen B tests followed between 1960 and 1963, looking at an accidental detonation in the event of a fire or crash. Such a mistake would not properly compress the radioactive core to create fission or fusion, but would lead to an unpredictable blast. Before each test the British produced a 'safety statement' for their hosts. For the first three years of Vixen trials these made no mention of the word 'plutonium'. For the last three it did, but by then it was no longer possible to pretend. The contamination was so bad that the commander at Maralinga had put pen to paper asking the senior health physicist about how, exactly, he was supposed to protect his men from the material left behind by these 'minor' events.

In February 1960, after being asked by the British how to proceed with Vixen B, Titterton said they should 'avoid formal communications on these contentious experiments and… proceed without going through the normal channels'. But by May the Australian politicians were growing suspicious, with Menzies withholding approval until he got the 'broad general particulars'. He finally gave the nod in September, and just three weeks later seven hydrogen-filled balloons which were supposed to be safely tethered, broke free and floated off. Two got halfway to the east coast, and were captured in New South Wales. An inquiry later found there was no safety plan, no mooring rules, no warnings, and that even the self-destruction device intended to detonate the balloons if they escaped didn't work. They had yet to be used in any tests; had they been, they could have carried radioactive contamination thousands of miles.

The Vixen experiments created the worst areas of contamination at Maralinga, littering the landscape with 22kg of plutonium-239, the same amount of uranium-235, 25kg of uranium-238, and 17.6kg of beryllium. Or at least, that was as much as was later found. To put these amounts into context, the average grain of sand weighs about 50 micrograms. If a human inhales an amount of plutonium equivalent to just two grains of sand - 100 micrograms - there is a near 100% chance of contracting cancer as a direct result.[70] Vixen B, alone, produced 22 *billion* micrograms of plutonium-239 - the equivalent of 22,000 bags of sugar. It's the same weight as a heavy goods lorry, or more sand than could be carried in the average dumper truck. And it was spread by the winds, and by man, across an area about the same size as Wales and Scotland combined.

Each Vixen B test had a 'feather bed', a massive steel structure set on a concrete pad. These were ruined by the blasts, and a new one was built each time. Debris from the previous beds - steel, lead bricks, paraffin wax, cables, concrete rubble, soil and rocks - was buried in pits. When they were investigated many years later, there were found to be 21 such pits containing an estimated 830 tons of material.

Plutonium, uranium, beryllium, polonium, cadmium, lead, actinium and scandium were also found scattered on the surface. While it was possible to find out how much radioactive material was used in each test, and subtract what was found on the ground, some remained unaccounted for. Although 101kg of beryllium had been used, and its toxicity well-known at the time, less than 2% was recovered from the soil. Another 20kg of plutonium also went astray. It was assumed - and hoped - to have been buried in the pits.

Britain sent one of its bona fide heroes to wade through the toxic detritus of Vixen. Howard Elliott lied about his age to sign up for the Second World War aged 17, eventually joining the Commandos. They were formed when Churchill asked for a mobile fighting force to carry out high-risk raids against Nazi positions in Europe.

He spent five months at Achnacarry Castle in the Highlands, training in live firing exercises, survival skills, explosives and hand-to-hand combat. It was the birthplace of the modern Special Forces, and inspiration for James Bond creator Ian Fleming, who commanded one of the units. After amphibious assaults and parachute training, Howard took on the Nazis by air, land, or sea.

He and his comrades in 6 Commando landed at Sword Beach in Normandy at 8.40am on D-Day on June 6, 1944, fighting their way through tank traps and into a swamp. Records show they took out four pillboxes and an artillery battery pounding the beach.

Howard's unit captured a bridge over the Aller, dismantled booby traps, then fixed bayonets in a charge at German defenders. When he was sent out on a motorbike for a minor task, Howard drove around a corner straight into a German convoy, which surrendered to him on the spot. And he told his family he once punched a senior officer to avoid being sent to North Africa, where he didn't fancy the conditions. 'He was a happy man, but quiet,' said his son Ashley. 'He was a very fit man, he could swim the entire length of a swimming pool underwater because of his Commando training. He didn't tell us a lot. We've had to put things together with his service records and our own research.'

After the war, Howard was posted to the RAF as an armament fitter, but retained a thirst for danger. He volunteered for chemical warfare experiments at Porton Down, and had a burn on his arm from exposure to mustard gas. Then in 1963 he was sent to Maralinga to take part in Vixen B. His job was to pick up lumps of plutonium from the ground where it had been blown. He died aged 58, after suffering for four years with colon cancer which spread throughout his body, and by the end had tumours erupting in his mouth.

After the war, when Churchill wanted a statue to honour the armed forces who saved Britain, Howard was chosen as the perfect example. He was the model for a memorial which still stands in the cloisters of Westminster Abbey. Ashley said: 'This country memorialised my dad for what he did in the Second World War, but dismissed the service he gave it afterwards, and which probably cost him his life. It makes me sad and angry that successive governments have done their best to sweep atomic testing consequences under the carpet... It was very dangerous work

Above: Howard Elliot, pictured back row, second from left, with comrades in Maralinga for Operation Vixen B in 1963. (Courtesy of Ashley Elliott)

Right: In 1946, Parachute Regiment corporal Howard Elliott was the model for the airman in sculptor Gilbert Ledward's Combined Services Memorial which still stands in the cloisters of Westminster Abbey. (Courtesy of Ashley Elliott)

and they were under-protected... I didn't really appreciate him enough when he was alive. When I was in school, I thought it was normal to be taught unarmed combat by your father. I would like to know he has been properly honoured for what he did.'

Alongside Howard was David Purse, a quiet, mild-mannered loner who emerged from his time at Vixen B with his health mostly intact. He had a far less glamorous military career, becoming a naval cadet in Anglesey after leaving school, but said he failed his interview to become an officer because he 'wasn't the son of a rear admiral'. Instead he switched to the air force, and was accepted into officer training at RAF Cranwell. Aged 29, he was a Flight Lieutenant posted at short notice to Maralinga, and looked forward to a warm country and a wild adventure. Instead he was put in charge of the airfield and its stores. It was David's responsibility to ensure the air and ground crews had all the spare parts they needed, that everything down to the toilet rolls was logged or stored, and kept in enough numbers to supply the ever-changing number of troops at work on the range.

He celebrated his 30th birthday at Maralinga, and took a keen interest in his surroundings when duties allowed - once posing with a pink galah on his shoulder, and taking hikes through the bush. He later told his son he was 'close enough to ground zero to see sand turn to glass' during the simulated 'accidents' taking place, with no protection. He said there was a rope with a 'do not enter' sign, and a wire fence, but the irradiated sand blew through it and onto food and faces.

David Purse, pictured with a pink galah on his shoulder during his tour of duty as an RAF sergeant at Maralinga, where a series of radioactive experiments were conducted. (Courtesy of Jacqueline Purse)

On his return home, he continued his habit of 'going walkabout', hiking through the Lake District with a couple of buddies. He was head of RAF mountain rescue before leaving the armed forces, and in 1970 he married Jacqueline, who he met while they both worked in a cattle food factory on the Wirral. A daughter, Kate, was born healthy two years later, and was swiftly followed by Steve.

'They wouldn't let me see him,' said Jacqueline. 'The nurses told me he was fine but they'd whisked him off to the children's hospital. I didn't know until I got home that Steve had a problem. And that was the first time David mentioned it. He said "I think it's my fault. It's to do with radiation when I was working in Australia." But there was nothing we could do about it, so we just got on with it.' Steve had been born with stunted limbs, a curved spine, and one leg bent 45 degrees the wrong way. As he grew up, he developed hydrocephaly, or fluid on the brain, which made him scream in agony. Doctors were unable to diagnose his problems. 'There are more than 200 forms of short stature in the world, and I don't fit a single one of them,' said Steve. 'They just call it an extremely rare genetic mutation. My dad never really wanted to talk about it, he was terrified of people finding out he'd been speaking about his service, and the Official Secrets Act, almost until the day he died. He only opened up in later life, and only to me, not my mum. He blamed himself for my condition, and he didn't want to talk about it with her.'

David developed severe skin problems on his lower arms and legs, which required huge tubs of cream to be applied every night before bed. His doctor took one look and said: 'You've lived your whole life in the tropics.' David replied no, just one year in Australia. The doctor said the skin problems were linked to radiation from the sun, and that somehow in those 12 months he'd received a lifetime's worth. It added to David's certainty that Steve's problems were his fault. 'It was only really years later, when an old RAF buddy got in touch and said there were a lot of children with problems, that he put two and two together,' said Steve. 'After that, he and I used to go on boys' trips, just the two of us. We would look at a map of the UK, say "we've never been THERE", and just pack up and go. Then it turned into a compass challenge, where we went to the most northerly, southerly, westerly and easterly points of the British Isles. We were going to visit the highest pub in England, Scotland and Wales, but we only managed one before he became too ill to continue.'

David developed Alzheimer's disease, and Jacqueline, who'd already had a lifetime of caring for her son, now cared for her husband, too. Kate had two boys, and although all were healthy, fretted over every headache and anguished cry, 'in case it was something nuclear', said Steve. Meanwhile he vowed to never have children of his own, for fear of passing on whatever genes may have been damaged. But in his late 40s he fell in love and got married, and in 2021 son Sascha was born. Steve's first question at every pregnancy scan, and after the birth, was whether the baby was all right. He worried about every cry, in case it was hydrocephaly. When Sascha was three he began saying 'Daddy, my teeth hurt' and a trip to the dentist found he had eroded enamel. He needed six baby teeth removed under general anaesthetic. 'He hasn't dodged the invisible bullets,' said Steve. 'We asked the dentist if it could be linked to my dad's service, and they said it was possible.'

Tooth enamel is known to be destroyed by radiotherapy, and it can also be damaged as a result of rare genetic mutations.

Today, Steve cannot straighten his arms, and needs a wheelchair, crutches, and adapted car to get around. Sometimes he uses an oxygen mask at night, because his lungs are underdeveloped. He works as an actor, with several radio series under his belt, but frets that as Sascha grows older, doctors won't have answers. He said: 'I know what it's like to have something go wrong, and doctors can only scratch their heads. I don't want that for my son, too. I know my dad blamed himself, but he was wrong. This wasn't an act of God, it was an act of government, and I want them to acknowledge they put men like my dad in harm's way and it affected their families.'

Grapple X, Y and Z cost the British government around £4.7m each, plus another £12m for the fissile material that had been needed - roughly equivalent to half a billion pounds today. In the aftermath, American co-operation was re-established, weapon systems were designed, and eventually the two nations entered an agreement whereby US missiles were used with British nuclear warheads.

Between 1962 and 1991 the allies exploded a further 24 devices underground in joint exercises in Nevada. These were tests on specific warhead designs, with comparatively low yields. The last took place under John Major's government, on

David Purse pictured with his son Steve, born with an undiagnosable genetic defect a decade after his father's service at Maralinga. (Courtesy of Steve Purse)

a refinement that enabled the person firing a missile to specify its precise yield, and therefore the circle of damage it would cause.

As the Cold War thawed and Ronald Reagan held summits with Mikhail Gorbachev, nuclear weapons began to be decommissioned all over the world. Today Britain retains one nuclear weapons system – Trident. Four nuclear submarines, constantly patrol the seas, each armed with 16 missiles and another 150 or so weapons stockpiled. Part British and part American, they are symbolic of how closely tied the two countries' nuclear fates now are. Despite having striven so long for them, Britain would never fire without US approval. While their threat is constant, it would in practice take days of safety protocols for the missiles to be made ready. What is now known as the Continuously At Sea Deterrent (CASD) has operated, without interruption, since 1969, keeping Britain and its people safe with a menace that never had to put its boots on. But in the same period, successive governments spent millions on a bitter war with the veterans who delivered it for them.

FALLOUT

THE MILPUDDIE FAMILY walked into the Marcoo crater one night in May 1957. Dad Charlie, pregnant mum Edie, their children Henry, who was around 11, and Rosie, about two, had set off with their dogs a year earlier from the Everard Ranges, 400 miles to the north. They were on their way to visit family at Ooldea, near the south coast, using a 'rockhole road' - a traditional track that ran north to south, linking the water sources and enabling long journeys to be made in stages. Knowledge of the paths was passed from parents to children among people who used these routes to stay in touch, despite the vast distances.

The barely-noticeable tracks which Penney had been assured were used only rarely were the Aboriginal equivalent of a motorway. The 'uninhabited, sterile ground' was dotted with fertile staging posts well-used by a nomadic population. And in the middle of this historic route was hundreds of square miles of freshly-irradiated Outback. Unbeknown to the Milpuddies, the tribe at Ooldea had been forcibly moved further south, to another mission at Yalata.

In April 1956, the AWRE declared that because Aboriginal people were usually naked and barefoot, or wore loincloths, they would be exposed to five times less radiation than Europeans in shirts and trousers. The logic behind this is unclear, and presumably this was based on the likelihood of fallout particles becoming lodged in clothing, but there was no suggestion troops should strip off to be safer. Even the poor safety standards of the time were more stringent for white people, than black.

The safety committee assured Menzies of 'continuous ground and air patrols', and a watch on 'the movements of the Aborigines' by native affairs officers and aeroplanes. But top brass were almost entirely ignorant of the Aboriginal way of life or how they might react. They thought a single check in the mornings was sufficient, 'because Aborigines sleep most of the afternoon and therefore would not move into a danger area overnight'.[1] Range Commander Colonel Richard Durance said the tribespeople feared 'evil spirits entering into them at night' and lit fires in the brush and jumped into the smoke 'to offset' it. Planes were told to look for these fires, on the rare occasions they looked at all. In fact, the Aboriginal people were hiding and putting their campfires out in order to confound searches by the blundering white men.

The Milpuddies did not speak a word of English, apart from Henry who had learned Bible quotations and hymns at a mission. Edie, speaking through a translator many years later, said: 'At a waterhole called Unguntju we heard an explosion and the earth seemed to be moving... We saw the soldiers and activity around Emu [Field]. The white fellow spoke to Henry. He had been to school... and he started to sing Jesus Loves Me and the soldiers heard that and understood. They told Henry that the mission was closed down; there was no water or food there. But we still went on.'

A few nights later, she said, the family 'were sitting camped in the bush. We had no water, just kapi [water] from the trees, knocking it out from the root... no pannikin [cup]... no clothes. When we lit a fire the white fellow saw the smoke and came in the morning.'[2]

Charlie, whose Aboriginal name was Tjanyindi, had unwittingly led his family into a designated 'dirty' area, either very close to or inside the crater. A patrol radioed it in, and their captain drove down in a Land Rover. He saw Charlie walking toward a military decontamination centre called Pom Pom, and stayed with him while sending troops to pick up the rest of the family. Their reports show the Milpuddies were terrified. They were manhandled, first into Land Rovers, then the building, and finally through showers. A local officer said: 'Showering was not part of the daily ritual... There was soap in there, and they had it in their eyes. The lubra [wife] was crying. I put them right under the shower and told them to keep their eyes shut very tight. I washed it out of their hair. I opened their eyes a little bit to let the water get the soap out of their eyes and got all the dust off them.'

What to the soldier was a kindness was to them an assault - pushed into heated water, scrubbed with unpleasant-smelling things, their eyes made to sting, and finally their eyelids forced open as they squeezed them shut. Charlie and Henry went through this process, but an officer ordered that no-one should touch the wife or daughter. After the showers, and to cover their nakedness, Charlie and Henry were made to put on trousers, and Edie was given underwear. They were photographed, and had radiation readings taken to ensure they were now within 'safe limits'. The family were put in a car and driven to Yalata, but 'it was a shocking trip down as they had never ridden in a vehicle before and vomited everywhere'. The manager at the mission said when they arrived the Milpuddies were 'in a state of apprehension and bewilderment and... they were frightened by the showering'. On the instruction of a civil servant from the Department of Supply, their dogs were shot in front of them as a precaution against contamination. A furious Charlie wanted to leave, but was not allowed.

No-one appears to have explained to the family what was happening. Edie was afraid, because of the large number of white men she had seen with no women around, that she had trespassed in a 'male' area. In her culture, such places were ceremonial, and taboo for females. No translator was sought. The whole family was left to suffer the physical and mental health problems which inevitably followed. Edie's baby was stillborn a short while later, and she believed 'the poison' at Maralinga had caused it. A son, Allan, was born in 1961 but died aged two from a brain tumour. A daughter called Annette died in 1965 at six months. There followed Sarah, born prematurely weighing just 2lbs, who went on to develop epilepsy. After the birth of their son Roger in 1968, Charlie left the family and lived alone. Friends said that 'he thought his eyes were sick because he was in the smoke. He finished with wife and talked silly. He camped self and went cranky'. He died in 1974, by then virtually blind, of heart failure.

Henry was diagnosed with advanced tuberculosis at 24 and later developed pneumonia, bronchial asthma and oedema, a build up of fluid in his tissues. Later, one of his daughters died in infancy from a heart problem, and another developed pneumonia and fluid in the eardrum. His sister Rosie, whose Aboriginal name was Milpadi, developed a heart condition. She lost a daughter in 1973, and a second daughter was born with heart problems and congenital dislocation of the hips. A third child, a son, was born with a club foot. The family's health was not checked until 1980, when these conditions were catalogued. Doctors reported that, regardless of any physical link to radiation, 'their experience and the (somewhat belated) concern about it has made Edie, Rosie and Henry confused and anxious... [they] have been subjected to a high degree of stress and unhappiness.' No-one, they said, had explained it to them in the intervening decades.

The fact 'natives' had been able to walk right up to the bomb site had the potential to derail the whole weapons programme. The Australians had been promised no humans would be allowed near. If that was untrue, public pressure would be too great to resist. Troops who knew about the Milpuddies were ordered to say nothing about what became known as 'the Pom Pom Incident'.

Between tests, Maralinga was under the command of the Australian army. Colonel Durance later admitted it was a 'political embarrassment' and he had made every effort to cover it up. 'We were all very guarded in anything that we had to say so that preferably it didn't get out from the range. I made this known to the men on the range that references to the incident were not to take place at all, particularly as they were under the Defence Special Undertakings Act and would remember that [they] could have great difficulties for them if they started breaking the security that was required of them in this matter,' he recalled.[3]

He reported it to the minister on the telephone, rather than on paper. The official report was classified 'Secret Atomic Guard' which limited readership to a handful of higher-ups. Even the patrol officer did his bit, telling the Milpuddies 'that as they had accidentally seen something of a white man's ceremony, they should not declare anything to other white men'. Reports were also made to the British government. A cable was sent from the UK High Commission to Downing Street on May 16, 1957, stating: 'The count was practically negligible on all four and it is felt that there will be no dangerous results. It is hoped to keep this incident from the press... all possible steps are, of course, being taken here to prevent this.'[4]

Yet nothing changed. Four months later Antler began, the veterans of which report the same 'safety' measures of wire fencing, English signage, and less than half of the servicemen got dose badges. Having two patrol officers made no difference to their practical ability to keep unauthorised visitors out of a vast area, criss-crossed by pathways white men couldn't see. Security concentrated on troops, who had to pass through barriers and have a good reason to go into firing zones. Those laying down the regulations seemed to think there was no risk of contaminated sand blowing around. The Kittens and Tims areas were eventually fenced off and guarded, but neither was capable of stopping the wind. Those who lived, ate, drank, and slept both near the forward area and further away at Maralinga Village recall the red dust getting everywhere, in bedding, clothes, food, and cups of tea.

The Minor Trials continued, with a total of 593 explosive experiments spraying radioactive material across previously-untouched parts of the firing range. Kittens involved 104 trigger detonations, producing about 120kg of natural and depleted uranium. The latter is now considered so toxic that its use in armour-piercing munitions is thought to have increased the death rate, and the number of birth defects, following the US invasion of Iraq in 1991. Tims involved 321 explosions over eight years, spreading 1.2kg of plutonium, 77kg of beryllium, 825kg or just under a ton of natural uranium, and 6,800kg - that's 6.6 *tons* - of uranium-238 across a wide area.

The 125 tests in Rats, combined with 31 Vixen A fires, led to 'significant amounts' of uranium being found for decades afterwards. A later report on Vixen B described 'jets of molten, burning plutonium extending hundreds of feet into the air. Following each trial, this plutonium was carried by the wind and deposited in long plumes extending large distances in the direction of the wind.'[5] The only thing that worked in the Brits' favour was that atomised radioactive particles are heavier, and denser, than a grain of sand, which meant depositions around the firing areas were recorded in clear 'plumes', in the direction of prevailing winds. But the same could not be said for every grain of sand that came into contact with these particles - each could become weakly irradiated, and blow where it liked.

Scientists knew radiation was deadly from the start. When x-rays were discovered in 1896, Nikolai Tesla noted beta burns to his fingers. 'Radioactive' medicines were touted as a quack cure until patients began dying, and they were taken off the shelves. In 1934, when Marie Curie died from aplastic anaemia after a lifetime studying uranium, her coffin was encased in lead. And in 1946 Hermann Joseph Muller was awarded the Nobel Prize for proving that there is no safe level for radiation below which genes will not mutate, and cause cancer.

In 1947, the Medical Research Council received a report from geneticist David Catcheside which stated: 'All quantitative experiments show that even the smallest doses of radiation produce a genetic effect, there being no threshold dose below which no genetic effect is induced.'[6] He said gene mutations, leading to 'sterility and gross structural changes' which could lead to the 'semi-sterility of offspring of irradiated individuals', were observable in viruses, bacteria, fruit flies, fungi and flowers. He said: 'It is very likely that the rates of mutation found to be characteristic for lower plants and animals will hold also for induced mutation in man.' An appendix listed the injuries likely to result in different organs exposed to radiation. In skin, it could cause blistering, ulcers, scars, cancer and pigmentation. In eyes, cataracts and burns. Bones could suffer cell death and cancer, changes to the production of white and red blood cells, and 'malignant overgrowth of blood'. Lungs could develop fibrosis and cancer, the genitals could become sterile or have reduced fertility, with 'hereditary changes in later generations due to chromosome and gene changes (in animals, not yet seen in man)'. Internal organs such as the thyroid, liver and intestines could lose their function, become scarred, or have cancerous changes. The lymph glands may have an overgrowth of tissue and a reduction in cells. Radiation could cause premature ageing to the entire body in general, and in cases of very high, one-off exposure, atomic bomb injuries could include 'radiation skin burns, acute radiation vomiting and sickness, haemorrhages and anaemia, sterility or miscarriages, prolonged wasting and death'.

With evidence from Hiroshima and Nagasaki, as well as experiments and leaks at Los Alamos, the harms were well-established by the time of Hurricane in 1952. Yet for it and each subsequent test, Penney and his team - like their counterparts in the US - set 'safety limits' for doses above which men should not be exposed. The implication was that a dose below that level was safe. In truth, they were just a way of minimising radiation, not making it harmless. Documents from 1951 show officials demanding insurance policies with payouts 10 times their usual salary for scientific staff in case any developed radiogenic diseases, and discussing the difficulty of recruiting scientists willing to work on the programme. The Hurricane naval task force commander, Admiral Arthur Torlesse, wrote a memo stating he would only order his men into danger 'in the knowledge that the Admiralty accept liability for those killed or injured on duty'.[7] The servicemen who took part in every successive nuclear test may not have known the risks they were facing, but those issuing the orders definitely did.

Before Hurricane, Dr W.G. Marley, head of health physics at the AWRE's sister agency the Atomic Energy Research Establishment, wrote an official memo confirming air crews flying through the mushroom clouds would be irradiated, but it 'is expected to be well below the lethal level at times later than a quarter of an hour after the blast'. He warned that 'the contamination of the aircraft is likely to cause radiation exposure quite as serious as that arising from direct exposure to the cloud'. But different information was passed from the British to the Royal Australian Air Force, which conducted many of the early sampling missions at Monte Bello and Emu Field. By that point, the advice read: 'The radioactive hazard to aircrews in flying through this cloud is negligible and there is no fear of the aircraft being contaminated.'[8]

In 1955 Prime Minister Anthony Eden was warned that if the scientists were to build an H-bomb, it could damage troops' DNA. A letter headed 10 Downing Street and dated November 14 says: 'The Prime Minister saw the report from Sir Harold Himsworth about the report of the Committee considering the genetic effects of Nuclear Radiation. His comment was: "A pity, but we cannot help it".'

The tower which held Mosaic II was only 100ft high, a tenth of what it should have been. The 'black mist' after Totem appears to have carried contamination along at ground level. Taranaki was detonated 1,000ft up, but if the heat radiation could vitrify the sand beneath, then sand could certainly be sucked up too, by simple convection. And Grapple Y, the biggest of all, took place at the end of the wet season.

The AWRE document on Eric Denson's experiment mentions 'the wet weather encountered during the trial'. Pasquini noted rain in the stratosphere, where it should not be. Veterans say the bomb dropped lower and closer than it should have done, and that 'black rain' fell on the island immediately afterwards. Among survivors, it is known as 'the bomb that went wrong'.

In 1956, there was proof about the global impact of nuclear testing. Three US weather researchers published maps in the journal *Science* using data from a new worldwide monitoring network. They showed the fallout from just one test - the 10 megaton Ivy Mike which was the first US hydrogen bomb in 1952 - had spread on trade winds. From one atoll in the Pacific, fission products travelled east across the entire continental US, north into China, and west over Asia and to Africa. It had crossed the Atlantic and headed towards the UK, and entered the upper troposphere above the Pacific, where it could circulate and fall just about anywhere.

The Campaign for Nuclear Disarmament held its first meeting in 1958, and there were widely-reported fears about strontium-90 being found in teeth and bones. After Grapple, the US and USSR agreed a partial testing ban, which Britain later joined. The world's three nuclear powers entered stasis. Harold Macmillan worked with both sides to formulate a permanent ban. The US and USSR held a successful summit at Camp David, and in 1960 France detonated its own nuclear weapon. Rather than derail the talks, it only made them more urgent.

But there was a stumbling block. New US president John F. Kennedy and Russian leader Nikita Krushchev couldn't agree on the details. Soon after JFK took office, the Soviets detonated the Tsar Bomba, 50 megatons of hellfire centred on an Arctic island. It was the most powerful weapon ever fired by man, and Kennedy had to respond. He moved Jupiter nuclear missiles to US bases in Italy and Turkey, and Khruschev reacted with a deal to put his own nukes in Cuba, just 1,500 miles off the coast of Florida. JFK sent a fleet to intercept, and for 13 days the world teetered on the edge of nuclear war. The belligerent posturing was dialled down in negotiations, and both sides agreed to withdraw their nukes from the other's back yard.

But even backing down from global conflict required a flourish, just to prove what was still possible. JFK decided the best option was further nuclear tests, but at a safe distance from his own people, and from the enemy. He asked Macmillan if he could use British troops and Christmas Island - which was in a better state of readiness, and politically more acceptable, than Nevada - for what became known as Operation Dominic. Codenamed Brigadoon in the UK, it involved 31 nuclear weapons being dropped from planes, floated under parachutes, fired from submarines, and detonated in the air, over the ocean, underwater and even in space. Seven were some distance away, but the remaining 24 were all dropped off Christmas Island, where 729 UK troops and hundreds more US servicemen were housed.

Among them was James Owen, known as Jesse to his friends, a 20-year-old Royal Navy sailor stationed at HMS *Resolution*, the shore base on the north coast. Over the space of 78 days at the end of 1961 and into 1962, he witnessed 24 blasts with a total yield of 23 megatons. He celebrated his 21st birthday between bombs

Left: 'Jesse' Owen witnessed 24 blasts in 78 days during Operation Dominic, and died 30 years later from his third heart attack. His family were compensated by the US government, but not the British. (Courtesy of Alan Owen)

Below: Jesse pictured in the garden with youngest son Alan in 1987, seven years before his death. (Courtesy of Alan Owen)

and on his return home, Jesse married and had three children. His daughter Laura was born blind in one eye, had a heart attack at 42, suffered repeated miscarriages, lost her teeth, and had three malignant cancers removed from her face. But she found out her dad was involved in the tests only after he died from his third massive heart attack, and her older brother Gordon died the same way just 18 months later.

Youngest son Alan, who believed all his life he had escaped the family illnesses, collapsed in 2022. Saved by CPR and a defibrillator that had been installed in the leisure centre where he was playing football at the time, he was rushed to hospital and kept on a ventilator, and in a medically-induced coma, for eight hours. His wife and son waited a further 20 hours to find out if he was brain damaged. When he came round, he was told his heart had stopped for eight minutes, and that he had a heart condition called hypertrophic cardiomyopathy. It causes the walls of the heart to thicken and harden, sometimes causing blockages and burst arteries. It can be caused by exposure to radiation, or genetic mutation. It is not known if the same condition killed his father and brother, as they were never tested for it. Alan had to have an internal defibrillator fitted, and took early retirement. Laura said: 'I was heavily pregnant when I found my dad dead on the floor. Now my younger brother Alan has been diagnosed with a heart condition, and only narrowly survived. You ask yourself if it's the radiation all the time.'

Some of the Dominic tests entered folklore, and sound implausible. Five bombs were detonated on the edge of space, for defensive tests to see if they could blow up incoming missiles. One, a 1,400 kiloton device codenamed Starfish Prime, lit up the Pacific sky, interfered with five orbiting satellites, knocked out streetlights in Hawaii and disrupted communications. Another, Bluegill Prime, blew up on the launchpad and heavily contaminated Johnston Atoll.

The official reason given for Dominic was 'weapon design and reliability', but from the variety of places and ways which the US chose to detonate its largest-ever series of nuclear tests, the sub-text was to show what it was capable of. Kennedy argued that the blasts were vital, to prove the need for a ban - a bit like demonstrating the need for gun control by shooting up a shopping mall. But it worked. The terrifying face-off forced everyone back to the table. Finally in 1963 the US, UK and USSR signed a deal forbidding any further explosions in the air, sea or outer space.

In a speech at the signing, JFK said: 'The number of children and grandchildren with cancer in their bones, with leukaemia in their blood, or with poison in their lungs might seem statistically small to some, in comparison with natural health hazards, but this is not a natural hazard and it is not a statistical issue. The loss of even one human life, or the malformation of even one baby, who may be born long after we are gone, should be of concern to us all. Our children are not merely statistics towards which we can be indifferent.' He is the only US president to ever acknowledge the genetic risks of the weapons trials, but to date no nation has admitted any culpability for the birth defects that were later reported by people living nearest to every nuclear testing site on Earth.

Testing still continued, underground. A total of 1,528 detonations have taken place beneath the Earth's surface, three times what happened in the atmosphere.[9] Those who live near the areas affected claim radiation has 'vented' through

Mushroom cloud of the Arkansas device, dropped by air on May 2, 1962, as part of Operation Dominic. A B-57 Canberra is in the foreground, with black smoke coming from its starting system, showing it was being prepared to take off and sample the cloud. (Wikimedia Commons)

geological fissures, and contaminated groundwater. After Dominic, the US and UK worked hand-in-hand on nuclear weaponry. The last joint test, codenamed Julin/Bristol, took place underground in Nevada in 1991.

The total weapons yield of the British nuclear tests in the Pacific alone was 7,069 kilotons, equivalent to 471 Hiroshimas. Including the joint efforts with the

US in Dominic, it was more than 2,564 Hiroshimas. In Australia the total was 217 kilotons, or 14 Hiroshimas, plus tons of radioactive material vaporised and spread across a vast area of Outback.

Once their testing programme had finished, the Brits cleared out their kit and equipment, and burned or dumped the rest. In Australia they handed over control of the range to local troops, and it became apparent that Maralinga was heavily contaminated. Some sites had been tidied - for example burials of the Vixen 'feather beds' - but only to the extent that was needed for troops and scientists to continue working there. Between Maralinga Village and the airfield was Building DC12, used by scientists as a laboratory. In it they kept a 'hot box' containing thorium-228, which was ultimately used for some of the Rats trials. There was a leak, and an extractor in the building pumped contamination up the building's chimney. Despite warnings in 1958, the box wasn't removed and buried until early 1960, in a clean-up known as Operation Ayres. When the building was checked, it was found to have had loose contamination scattered throughout, where the scientists had been working. A year later, Ayres 2 removed and buried a second hot box, and some of the Vixen B leavings. No-one seems to have wondered what happened to the thorium that went up the chimney.

Records show[10] that contaminated soil, and three tons of 'contaminated material from the forward area' including cables were likewise given 'radioactive burials' in Maralinga. But that wasn't all. In with it went 282 steel drums, each capable of holding 44-gallons of decontamination sludge, fluid and solids that had been removed from the Canberra 'sniff' planes after sampling missions at Grapple. The planes had flown through the clouds, landed and been 'cleaned', before flying back to base at Edinburgh Field in South Australia. Once there, they needed decontaminating yet again, and it was the resultant gloop of water, solvents and engine grease captured in storage tanks that now had to be disposed of. When photographs were taken of the radioactive 'Maralinga cemetery' in 1975, there was 'a complete lack of herbage'.

In mid-1963 the Australian officer in charge of the range organised another effort, Operation Clean-Up, every Tuesday afternoon, with all troops required to take part. They removed 175 tons of contamination from the Minor Trials sites, and buried it. Some sites were left alone. At each firing pad several hundred yards of Outback was peppered with plutonium fragments, and what to do with them was a quandary.

In a sign of how much had been kept in the dark, the Australians felt compelled to write to AWRE asking exactly what had been left where. Whatever the answer, in 1964 the Australians decided to try again. The next clean-up, Operation Hercules, aimed to clean what they could, and bury the rest under concrete, while exhuming and reburying the previous efforts. Plutonium from the Minor Trials presented the biggest headache - too small to pick up, too dangerous to be left lying around, and still resolutely to be found in the top 1cm of soil. The solution arrived at was to plough it in, to a depth of 15cm. Cobalt pellets and scientific instruments were buried.

But a survey the following year found the area was still contaminated. In 1966, the British announced they no longer needed Maralinga, and offered to conduct further ploughing, covering the burial pits with concrete, and sealing up drains, as 'a full and final settlement of all obligations whatsoever'. The Australians wondered how safe it really needed to be - after all, the Aboriginals had gone, and the troops weren't coming back. Unbeknown to those making the decisions, indigenous people were already returning, thanks to the missionaries at Yalata who drove them in and out for ceremonies. When Operation Brumby was launched in 1967, plutonium that could still be found on the surface was collected again, and buried in 22 pits covered in 650 tons of concrete, surrounded by fencing. The ground was re-ploughed. It was found that the area used for Vixen A was particularly hot, and the topsoil there was scooped up and dumped in the Marcoo crater, along with contaminated aircraft and caravans. They were set alight, and the resultant charred remains covered with about five feet of soil. More cobalt pellets from the Tadje blast were collected in drums and buried near Maralinga Village, along with a 'massive amount' of natural uranium, sealed in lead. Each major test site got an extra covering of soil. Other areas were ploughed, scraped, and graded to mix contamination with uncontaminated dirt.

None of this was ideal. Burials by British troops were intentionally unmapped in the hope they would not be found, but in reality it meant they could not be checked, maintained, or counted. Ploughing made it harder for any of the contamination to be removed in future, and spread radioactivity across a wider area. A report by scientist Noah Pearce, which was leaked to the *National Times of Australia* many years later, said tractors were used to churn plutonium into the topsoil over an area of 500 acres. Nevertheless, in 1968 the land was formally handed back to the Australians, with a joint declaration that the Brits 'have completed decontamination and debris clearance... [and] are released from all liabilities and responsibilities'.

By 1972, wind erosion had taken much of the topsoil away, destroying vegetation and exposing concrete, metal and glass-like material. Plutonium was found, and removed, again in 1979, but five years later yet more had to be fenced off. There had been six 'clean-ups' at Maralinga, and it still wasn't clean.

At Christmas Island, a similar process was underway. In 1963 the AWRE sent scientist Alec Oldbury to complete 'a final radiological survey' before formally abandoning the base. He and one assistant had the job of sampling, monitoring, fencing and signposting the areas used for testing, plus examining laboratories and airfields. They relied on Gilbertese islanders for manual labour, supplemented by the few remaining British troops. Oldbury's subsequent report says that 'a tolerance level' of 'four counts per second' was 'taken as the clean/dirty line' of radiation.[11] But international atomic safety regulations at the time, and still in force today, define a reading of 4 cps to be radioactive contamination; the line at which Oldbury was declaring things 'clean' was, and is still, considered dirty.

His report shows a water tanker contaminated on the inside (400 cps), the concrete pad where sniff planes had been washed down (up to 350 cps), and the pond which collected the water from the camp laundry and decontamination areas (up to 200 cps). The tanker was buried at sea; the concrete pad sealed with bitumen; and the pond liner sliced up, placed in drums, and dumped overboard.

Jeremy LeBois, chairperson of the Maralinga Tjarutja, examining a buried steel container being exposed by weather erosion. No-one has any idea what it contains, its purpose, or whether it still presents a danger. (Courtesy of Daniel Everitt-Lock, True Perspective Films)

They considered smashing up the concrete and throwing that over the reef as well, but decided it was too difficult. A decontamination centre built by the Americans was razed to the ground, while drying machines and a washing machine from the camp laundry were taken out to sea after they were found to be contaminated. The galvanised tanks used to store laundry wastewater went with them, along with sections of pipe, drums of caustic soda and sulphuric acid. The area chosen for the disposal was three miles out, where the sea bed was more than a mile down. So far as anyone knows, it's still there.

Oldbury found the areas used to park planes had patches of concentrated radiation. He theorised that the bitumen, in tropical rainstorms, was being broken up, and forming small depressions in which particles collected in an atomic puddle. He ordered his men to clean out the dips, and the counts dropped, but his report says that the entire area would need to be relaid, which was beyond his abilities. He complained that many of the AWRE buildings had been broken into and looted, a mobile caravan was missing, and generators had been raided for parts. He wrote that 'particular annoyance was felt regarding the disappearance of a small amount of protective clothing, electric fan, electric kettle and Thermos container' which he'd carefully stashed on a previous visit. Oldbury said there was no point 'trying to find the culprits', but resentfully noted that 'during the course of the operation casual observation revealed some of the missing items in the possession of Gilbertese villagers and service personnel'.

He resurveyed the contaminated areas, and decided the counts were low enough to be safe. Then, 'in view of the survey results, all warning and other notices that might have given some indication of the work carried out in this area were destroyed by burning'. But there remained 'vast, untidy and unsightly dumps' of military equipment, and he said it would be best to continue to avoid the water tanker, pond, wash down pad and other areas.

Robert McCann was there at the same time. As a 19-year-old in the Royal Engineers, he had seen six bombs during Grapple. He operated stone crushers in a quarry, drove a drilling machine, and built fuel tanks. He said: 'We had no protection. We were given a pair of First World War sunglasses for the first test, and for later ones were just told to put our hands over our eyes. After Grapple Y, the biggie, it rained afterwards. Not real rain, more a river from the sky. And if that didn't get you, they sprayed toxic DDT over the island almost every day to kill the insects. They sprayed it over us, too.'

Six years later, he was ordered back to help clear up after Dominic. He said: 'They left such a mess, it was terrible. Trucks, bulldozers, barrels. We were ordered to put it all on an old landing craft, take it over the reef and scuttle it. It was my job to put the explosives on. Goodness knows where all that stuff ended up, or how contaminated it was.' He later became a schoolmaster and deep-sea diving instructor, married and had two 'mercifully healthy' sons. He developed prostate cancer in his 50s, and aged 85 found a growth on one buttock. After repeated badgering of doctors, it was biopsied and pronounced to be a myxosarcoma, a

Robert McCann, on the truck, pictured with his fellow sappers on Christmas Island during the clean-up in 1964. Note the protective clothing. (Courtesy of Robert McCann)

very rare cancer of the soft tissue known to be radiogenic. A fist-sized tumour was removed, leaving an 18-inch scar and not much left to sit on.

'The surgeon told me the only other cases she'd seen were on people who'd been exposed to radiation, and thinks it must be due to Christmas Island,' said Robert. 'I loved the posting at the time, it was brilliant apart from the bombs. Some of us had it bad. I know of one chap who was a pilot in the planes they ordered through the cloud, and he died soon after he got home. His widow told me all his skin had fallen off.'

In 1969, the last British troops left, taking with them whatever was re-usable and abandoning the rest. Six years later US scientists surveyed Christmas and the other atolls it had used for weapons tests, and found 11 fallout nuclides, including caesium-137, which was in more than 20% of samples. The highest levels were found in leaves of trees near the airfield, and in a fish caught off the reef. It considered these 'trace quantities' with no cause for concern.

In 1979 Kiribati - the collection of islands of which Christmas was one - gained independence, and began agitating for new checks. In 1981 New Zealand sent another team, paid for by the Brits, which found radioactivity in soil was 'consistent with global levels', with no risk to islanders. But the report relied on Oldbury for background information, and 'estimated' exposure rates by 'making some assumptions' about what was deposited from the weapons. It had no information from AWRE about what was in the bombs, or where it may have fallen.

They sampled coconuts, soil, rainwater, drinking water, fish, crabs and the island's lagoons, and although they found caesium, plutonium and strontium in much of it, levels were low. Their report concluded: 'No radioactive contamination was detected which would present a hazard to resident islanders.' It did not discuss the possibility that, in a place without many of the background sources of radiation which affect parts of Britain, even small amounts might produce a marked effect.

No documents uncovered so far mention what all the contamination they found meant for the men, and indigenous people, who had lived and served there. If it was necessary to encase, bury, or dump at sea items like gloves, furnishings, laboratory equipment, engine grease, washing machines, and even the things which held water to drink or wash with, then surely the men that had used all those items on a daily basis - sometimes for years - could not possibly have been safe, much less any locals who were there at the time, or later liberated leftovers for their own use.

The fallout followed the veterans home. In 1958, Eric Denson had left Christmas Island early. Shirley received a telegram from his commanding officer saying he was being sent home 'for operational reasons' as he had 'exceeded his limit'. Before the tests he was described as 'confident, kind, unflappable'. After he returned home he developed mood swings, depression, and crippling headaches. They were obvious the day he got back.

Shirley said: 'He was not himself; he was pale, he'd lost weight, and looked like he hadn't slept for days. He was a reserved man, never said anything unnecessary, but that night he talked non-stop – about God, sunsets, his childhood, what it was like to fly. And I saw he had a terrible rash on his chest, from his neck to his waistline, and covered with tiny blisters. I'd never seen anything like it before, but of course now I know it was a radiation burn.'

Shirley gave birth to three more children, all girls. Eric's mental health deteriorated: he became moody, drank heavily, and at one point told her he felt his brain was under a suffocating black cloud. She remembered: 'He knew something was happening to his head, but he didn't know what. He would hold his head with tears running down his face. He was in agony. We took him to psychiatrists but they had no idea what was wrong.' In 1968, a decade after the tests, they decided it was depression.

Eric became one of the youngest squadron leaders in the RAF, but his problems were noticeable and he was 'persuaded' to leave. He found work as a commercial pilot and started to cut his arms, wearing long-sleeved shirts to cover the scars. Sometimes he stubbed cigarettes out on himself. Twice, Shirley found and saved her husband as he tried to take his own life. One night she found him sharpening an axe in the kitchen, a glass of whisky by his side, preparing to kill her and the children before himself. She sat beside him and talked him round, convincing him to go to bed. The third time, on July 8, 1976, he succeeded in taking his own life, slashing his wrists in a wood near their home. Shirley was taken by police, along with one of her daughters, to identify his body as it lay at the foot of the tree he had chosen to rest under. By his side was a bloodstained note saying he couldn't 'take it any more'. He was 44.

Shirley was left a homeless widow with four daughters, one with developmental abnormalities. She found a home through a military charity and went out to work, but as her children grew up to have families of their own she began to notice a pattern. Three grandchildren have unexplained spinal problems. A granddaughter had three adult teeth missing, and a great grandson had three sets of teeth. A great-granddaughter was born with an extra tube between her kidney and her bladder. It was not properly connected, and as a baby she needed antibiotics to keep peritonitis at bay. Another granddaughter and her four children all suffered excruciating joint pain. When one great grandson got adult teeth, they were so filled with holes he had to have four removed. More than a third of Eric's descendants have a congenital birth defect: the normal rate in England is 2%.

In 1980 the festering injustice finally broke into a scandal. Yami Lester made front page news in Australia, and at veterans' reunions in the US and UK, the passage of 30 years had put things into stark perspective. Former servicemen were hearing of comrades dead before their time from rare conditions, of families scarred by both birth defects and worry. A report was ordered from the Australian Ionising Radiation Advisory Council (AIRAC) which found the genetic fears of veteran families were 'unfounded'. It did little more than restate the British position, that there were stringent safety policies in place and they were followed to the letter. It found, for example, that the Milpuddies had only a tiny dose, and it 'cannot possibly have led to any identifiable ill-effect'.[12] It added the black mist probably didn't happen, and 'the precautions taken to ensure that Aboriginals

Aboriginal tribal elder Yami Lester broke the scandal open in 1980 when he first told the story of the 'black mist' from Operation Totem which blinded him, and killed many of his family. (Wikimedia Commons)

were not harmed by the tests were adequate and effective'. It did little to quell the noise.

At the end of 1982, a BBC *Nationwide* documentary highlighted the issue, and in January 1983, the first widow came forward. Ellen Grigsby's husband Bill died from a brain tumour after serving at Maralinga, where he had to search for Aboriginal people, and collect ground samples. She told Channel 4 documentary *Broadside* that a pathologist found 'nuclear debris' in his brain after his death in 1977, but she was still refused a war pension.

Investigative reporters smelled a scandal, and the cause was taken up by Joan Smith in the *Sunday Times*, and David Leigh and Paul Lashmar at *The Observer*. Joan later wrote: 'The articles attracted many letters; some were from men who wanted to connect to a veterans' organisation in Britain, some from people who just wanted to tell someone their experiences. One man wrote from Devon to say how pleased he was that the claims of the veterans were being taken seriously.' She described one man who wrote to her about a skin irritation he'd developed aged 22, soon after the tests, and which had plagued him since. 'I have often wondered if any of my ailments over the years had anything to do with the H-bomb test. I even asked a doctor and I was scoffed at,' he told her. 'I used to wear white cotton gloves when I went to bed but this didn't stop me from scratching my face and chest until they bled.'

The scandal was becoming unmanageable. Margaret Thatcher announced that, while she had been assured no-one was harmed, her government would investigate. She told the Commons: 'The MoD has announced that it will conduct a health study of all personnel who were serving in connection with those tests in Australia and the south Pacific. That is a survey of about 15,000 personnel to see what the effects have been. That is the best way to go about this problem.'[13] Veterans feared it would be a whitewash, as it later turned out the study would not consider their medical conditions, miscarriages or birth defects, but just their death rates. At the same time, Australian veterans were threatening court action against their own government.

In March 1984, as the drumbeat of scandal intensified, the Australian and UK governments were in regular contact. One memo from the assistant secretary at the Australian High Commission in London to a counterpart at the Department of Resources and Energy in Canberra included a list of radiation doses received by the Buffalo indoctrinees, along with the incredible claim that AWRE had only 'discovered' them the day before. It detailed collusion between the two governments to ensure that the Australian Freedom of Information Act - the UK did not have one at the time - could not be exploited by veterans. It added: 'I am advised by Mr McTaggart of the Ministry of Defence that he is preparing an MoD policy guidelines paper on the release of information to the public relating to the tests... MoD are concerned that their nuclear veterans association gain access to documents via their Australian counterparts... I have indicated there are certain provisions in the act which could be used to restrict the release of such information as appropriate.'[14]

Such discussions between governments were not known publicly, and it is not clear how much those in power knew, either. In 1984, junior defence minister Adam Butler wrote a letter to MPs, saying only 50 personnel had a 'planned special exposure level' because they collected samples. He added: 'All the evidence I have indicates that proper precautions were taken to safeguard the health and safety of those involved... it is important to note that the standards in use at the time were comparable with those of today.' Comparable, but not in a good way. In 1950 the international radiation protection standards for medics was for a maximum exposure of 3 mSv a week. By 1984, when the letter was written, it had been reduced to 0.9 mSv a week. Butler was either lying, or had been misled by his officials.

A few months later a fiery Scottish ex-sapper called Ken McGinley, who ran a bed and breakfast with his wife Alice in Dunoon, wrote to the letters page of the *Daily Record*, asking if anyone else suffered health problems after the nuclear tests. Dozens came forward, and the British Nuclear Test Veterans Association was formed, with 500 members in the first six months. That July, Liverpool MP David Alton told Parliament that Australian officials had published accounts of Operation Buffalo, telling the Commons: 'British servicemen were used as atom bomb guinea pigs, and that is profoundly disturbing. If true, the government has a duty to make known the names of the military personnel who issued such inhumane and immoral orders,' he said. He demanded a public inquiry, and the official numbers for those involved jumped to 20,000.

Grapple survivor Ken McGinley, pictured at home in Johnstone in early 2024, became the charismatic leader of a recognition campaign with the motto 'All we seek is truth and justice'. He founded the British Nuclear Test Veterans Association, and overturned a law forbidding servicemen from suing the government. Adoptive dad to Louise and husband of his beloved Alice, he died from lung and kidney cancer in June 2024. (Courtesy of Professor Chris Hill)

The BNTVA motto was 'All we seek is truth and justice', and by November it had reached the ear of journalist Paul Foot, a renowned campaigner and investigative reporter who had a weekly column in the *Daily Mirror*. In November 1983 he wrote about 'The Curse of Christmas Island', detailing how RAF veteran Ian Loudon, who had served at Buffalo, had died of cancer aged just 46.

In Australia, the political situation was even more febrile. Not only was that country the one irradiated, and had its own issues about mistreated indigenous people, but their Prime Minister was not on friendly terms with his British counterpart. Bob Hawke was a blunt-speaking Labour politician who declared he had a 'love-hate relationship' with Thatcher. The two were at loggerheads on race relations and disarmament, with Hawke wanting to declare the Pacific a 'nuclear-free zone'. In 1984 he authorised a Royal Commission into the British nuclear tests, a public inquiry that was calculated not just to annoy former overlords, but to curry favour with the electorate by providing them with a blameable bogeyman. It was chaired by Jim McLelland, a judge who had also been a Labour politician and served as a government minister.

The hearings, held over more than a year in Australia and the UK, were a catastrophe for the British. Veteran after veteran took the stand and revealed the negligence with which they and the land had been treated, while media informed the public for the first time about the toxicity of the Minor Trials, and the size of the blasts. Scientists gave evidence under oath, and Penney was a star witness, reluctantly forced to account for every aspect of the testing programme which, to be fair, he had not always been personally involved with. The UK was compelled to declassify thousands of documents, and many were seized on by British veterans. Those detailing early fears about genetic hazards, plans to irradiate men and equipment to see how they functioned, and the horrifying experiments of Buffalo and Mosaic, all became headline news.

And for the first time, the world heard the testimony of those people who were most sidelined. The Pom Pom incident was revealed, and the Milpuddies gave evidence, along with Almerta Lander, Lalli Lennon, and Yami Lester. There were revelations about Australian servicemen given less protective equipment, of British officers 'talking down to the colonials', and how scientists circumvented the Australian safety committee with the connivance of British-born Titterton. Troops spoke of repairing broken-down vehicles in the forward area, while their rubber respirators had to be discarded after becoming clogged with sweat. Some claimed to have been sent through decontamination units, only to be ordered back out to repair, use or move equipment that had not been washed down at all. There was evidence about scientific negligence, military incompetence, and flatly-racist treatment. Some veterans were insistent all was well, while others spoke of missing medical records and subservience to a foreign power. The Australian government had accepted without question the British refusal to share technical data, did not demand the information that would help to understand the risks, and aided and abetted them in pulling the wool over the eyes of the public.

Its unhappy conclusions were that Monte Bello was not the right place for Hurricane or Mosaic; that Totem was fired when the wind would produce 'unacceptable levels of fallout'; that the crew of the Atomic Tank had been exposed; that air crew had been negligently ordered into the clouds; that Aboriginal people in the vicinity of every test from Hurricane to Vixen were either ignored or inadequately cared for; and that the use of cobalt-60 at Tadje was discovered only by accident and led to exposure. But it ruled that the safety policy was 'reasonable and compatible with international standards' of the day, while at the same time finding the 'safety measures' were not adequate. It found *Diana*'s crew were mostly protected from radiation, that there was no proof Yami Lester's blindness was caused by the black mist, that genetic impacts on families had not been proven, and that 'the level of radiation [at Maralinga] will decay to one of no significance during the lifetimes of the young people now returning to the area'.[15]

Yet it also, counter-intuitively, demanded further clean-ups, and stated that contamination in some areas was 'unacceptable'. It said the AIRAC report was flat wrong, and it should never have concluded that safety standards had been followed or that no-one but a few air crew were exposed. It said: 'The Minor Trials were not minor at all in terms of their consequences… the information relayed about them from the British to the Australian governments was notable for its economy.' On Buffalo, it found: 'The attempts to ensure Aboriginal safety during the Buffalo series demonstrate ignorance, incompetence, and cynicism on the part of those responsible... the inescapable conclusion is that if Aborigines were not injured or killed as a result of the explosions, this was a matter of luck rather than adequate organisation, management and resources allocated to ensuring safety.'

It concluded that the Australian tests were 'a drama characterised by persistent deception and paranoid secrecy' and added: 'In their desire to avoid international repercussions, the British authorities embarked on a course of determined concealment of information from the Australian government, aided and abetted by the "Australian custodian", Titterton.' It ruled that Vixen should never have been allowed, and most trials took place in contravention of the rules imposed.

The commission made seven recommendations, including an immediate clean-up at Maralinga and Emu, their eventual return to Aboriginal owners, and the collection of radioactive debris - those metal lumps of HMS *Plym* which had rained down after Hurricane - from Monte Bello, with all costs to be borne by the UK. It also demanded the Australian government compensate the indigenous people affected.

Long-term polling shows the Australian public mood shifted towards republicanism in the wake of the hearings. It also contributed to Hawke's re-election: the commission's findings conveniently pointed the blame at Menzies. He had not only led the conservative Liberal Party, but had died seven years earlier and was unable to defend himself. Hawke could paint his party as the saviour of veterans and indigenous people, and if Britain was slow to cough up compensation he could point the finger of blame at his nemesis, Thatcher.

Veterans in Britain felt that their own case grew stronger as a result. With all the media attention came lawyers offering to help, and in 1985 ex-sapper Melvyn Pearce, managed to convince the House of Lords the government could no longer rely on Crown immunity to stop it from being sued for harm done to its troops. It led to the repeal of the Crown Proceedings Act in 1987, and as a direct result, thousands of veterans since have been able to sue for injuries sustained after taking the Queen's shilling - an outcome that has cost the MoD billions, but has yet to bring automatic compensation for survivors of the nuclear tests.

A year later, the NRPB study revealed its findings. The 7-page report in the *British Medical Journal* was pored over by the journalists and politicians, who read - to their astonishment, considering the stories veterans had told and the findings of the Royal Commission - that they were no more likely to die than anyone else. The study had found the total number of men who had served at the tests was now 22,347. It compared them to a similar group of ex-service personnel, matched for age and service, who had served elsewhere at the same times. It looked at 38 different causes of death, including coronary heart disease, alcoholism, bronchitis, cancers, car accidents, drowning and suicide. The tables it reproduced showed just 1,149

One of dozens of memorials to the nuclear veterans, paid for by survivors and families themselves via the BNTVA, and installed in civic spaces all over Britain. They are, to date, the only public symbol of what was the UK's biggest military operation since D-Day. (Wikimedia Commons)

deaths among the veterans, and only 134 with a tumour of some sort. It broke down the types of cancer that had killed test veterans - tongue, mouth and pharynx (8), stomach (26), liver (12), bladder (10) and so on. It found 27 had died from cancer of the lymph glands, and 15 from leukaemia. Overall they died at a similar age as the controls, but the higher number of blood and bone cancers, though small, was statistically significant. The report concluded: ‘Participation in the nuclear weapons test programme did not have a detectable effect on the participants’ expectation of life or on their total risk of developing cancer, apart from a possible effect on the risks of developing multiple myeloma and leukaemia.’[16]

In the five years since forming the BNTVA, McGinley had been forced to sell his guesthouse when the American navy, which had a submarine base nearby, blacklisted it. The day before the report was published, he told *The Scotsman*: ‘It’s about time the British government and the MoD were taken by the scruff of the neck. They know they have left a terrible legacy behind them with the nuclear tests, and they are the people who are answerable.’

Journalists were given a copy of the report the day before it was published in the *BMJ*, and faithfully reported its findings that, although there was a slight increase in some cancers, the veterans’ life expectancy was no different. McGinley was furious, and claimed the figures had been massaged somehow. The MoD said it would pay war pensions to any test veteran with the right sort of cancer, and most of Fleet Street moved on to other things. After almost a decade of public outcry, the story seemed to have run its course.

Just a few months later, US president Ronald Reagan signed the Radiation-Exposed Veterans Compensation Act, providing disability benefits for those who had served in Japan after the war, internees affected by exposure, or veterans of its own nuclear test programme. It included a list of 21 radiogenic conditions which would lead to a payout. Even though it was theoretically possible they’d develop such an illness anyway, it was politically unsupportable for a US government to argue the point: it had irradiated its own land, its own citizens, and ex-troops were revered by the public. Reagan called them ‘a truly unique group of veterans’, and said: ‘The nation is grateful for their special service, and [this act] makes clear the nation’s continuing concern for their welfare.’[17] Two years later, in an update of the legislation, the right to claim was extended to British troops serving at Operation Dominic, who could get $75,000 if they developed one of the 21 conditions. The UK government, usually so keen to follow the Americans in every defence policy, saw no reason to do the same for its nuclear survivors. They were no longer politically-imperative, and as the media lost interest, so did voters and politicians. Without an instruction from ministers to alter its course, the ship of state sailed on, regardless.

Unable to disprove what a government was saying with all its might, McGinley and his comrades pressed on blindly. In 1989 the Official Secrets Act was updated, freeing veterans to speak out fully for the first time, and four years later McGinley and another veteran, Edward Egan, launched legal action claiming their human rights were breached. The case went all the way to the Court of Human Rights in Strasbourg, where it ultimately failed. Both men argued they had been denied their medical records. The MoD insisted it had provided all that was available, but McGinley's mental health would be damaged if he were allowed to see all of his. He was given a summary, with the full version provided to his lawyers. The court decided that any suggestion the state was holding back dose records of any sort was 'speculation'.[18]

In Parliament the scandal split along party divides - with Tory government ministers continuing Thatcher's and Churchill's line that nothing was amiss, and opposition MPs insisting it was a scandal. In 1990 the Labour MP Bob Clay proposed an amendment that would have given a war pension to all veterans who developed a radiogenic cancer. The list could have included leukaemia, bone, bladder, lung, thyroid and brain tumours. The Labour Party ordered all its MPs to back the amendment, although with the government voting against, it was always going to be defeated.

Archie Ross was typical of the nuclear veterans - proud to have served his country, and unwilling to believe anything went wrong. An RAF corporal who

Above left: RAF corporal Archie Ross helped dozens of widows apply for war pensions after finally realising that his family's health problems could be linked to radiation exposure. (Courtesy of Stuart Ross)

Above right: Archie, his wife, and children Tracey, Stuart and on the right Julie, whose disfigured arm can just be seen. The limb has grown throughout her life, causing further health problems. (Courtesy of Stuart Ross)

flew provisions to the Grapple task force, he happened to be on Christmas Island for Grapple Y, and like others was ordered to sit in the open, covering his eyes, and then turn to watch the blast. He later married and had three children. Tracey and Stuart seemed healthy, but daughter Julie was born with a right arm and hand many times their usual size, and a twisted rib cage.

As she endured dozens of surgeries, Archie put it down to bad luck. In the 1970s he discovered a layer of skin growing between his eyelid and his eyeball, a condition which is usually only seen in elderly people. Then Tracey had a son with Down's syndrome, and he realised his whole family may have been genetically damaged by radiation. He set up the West Midlands branch of the BNTVA, and was renowned for helping dozens of widows through the process of claiming war pensions. He died in September 2015, having been diagnosed first with a severe heart condition and then non-Hodgkin's lymphoma, both of which are common among veterans. One of his last acts was to sign a statement for yet another war pension hearing, but as death approached this normally cheerful man became bitter at the continuing failure of the country he loved to recognise what it had done.

His widow Christine said: 'At first he wouldn't have any of it – he always said the government had looked after them and wouldn't have let them do anything that would harm them. But after he rang a colleague – who had a similar eye condition – his view changed... It was something that grew year on year. It was daunting and he had so many knock backs, so many slaps in the face but he never gave up. There would be times I would have given up, when a lesser man would have given up but not Archie – he never did. He was a very special man.'

In 1993 the NRPB study was repeated, and this time found no increase of leukaemia. The war pensions were quietly rescinded. When Labour came to power four years later, 38 of the MPs who had signed the Clay amendment were government ministers. Tony Blair had voted for it, along with defence secretary George Robertson, pensions secretary Alastair Darling, and 14 others in the Cabinet. The BNTVA was hopeful that they would keep their word. But Blair now backtracked. 'I am bound to repeat that there is no evidence of excessive illness or mortality amongst all of the veterans as a group that could be linked to exposure to radiation as a result of their participation in the tests,' he said in a letter to BNTVA secretary Sheila Gray. 'You fairly draw attention to my support for Bob Clay's amendment back in 1990. However, I have to say that, on the basis of full access to all relevant documentation, much of it now in the public domain, it is clear that very few of those present at the tests would have received a measurable radiation dose.'[19] Records show the line was drafted for him by a parliamentary clerk within the MoD, without reference to any of the documents in question.

The veterans felt betrayed, and the BNTVA kept pushing. Sheila insisted: 'We know that our men are being harmed, we know that our children have been genetically damaged and our grandchildren have been damaged. I'll give you one astounding fact. In the normal population, one child in 100,000 has adrenal cancer - in our group it is four in 2,500. You can see the similarities with the Chernobyl men, with the New Zealand men, the Fijians. But New Zealand and the US have both accepted it and are paying pensions to their test veterans.' She organised a mass-letter writing effort from veterans to their MPs, but archives show the MoD fought back with a draft letter from Defence Secretary John Spellar, stating the veterans' 'task was to provide logistic support (mainly construction work), and most certainly not to be subjects of the tests'.[20]

The BNTVA thought it had found a solution in 1998, when the University of Dundee published a study based on a questionnaire sent to 1,041 veteran families by researcher Sue Rabbit Roff. She wrote: 'Thirty per cent of the men in this sample have already died, mostly in their fifties. Two-thirds of them died from cancers that are pensionable in the US as presumptively radiogenic.' She described a catalogue of health problems that focused on the spine, skin and gut. She found twice as many cases of multiple myeloma as were in the NRPB study, and that one in seven men had never fathered children at all.

As much as this research chimed with stories that had been told in the media by veterans and indigenous people, the MoD found it easy to refute. These were the self-reported claims of a group of individuals who had joined a campaigning organisation, it said, so the sample was biased. There were no 'controls' to compare them with, and besides - the number of veterans were only a fraction of the 22,000 in its own study.

But the government research had never looked at what happened to children, and there Rabbit Roff broke new ground. Of 2,261 children in her survey, almost half had a health problem. More than 50 had died as infants, about double the expected rate, and five children had cataracts, a condition rarely seen in youngsters. There were reports of children with extra and missing teeth, hair loss, 46 who had a heart problem and 40 with cancer. Among 2,342 grandchildren, there was three times the expected rate of spina bifida, three leukaemias and 11 with a hip deformity. Rabbit Roff said the rates of spina bifida in descendants were about five times the national rate for live births. She noted cases of skeletal abnormalities, short stature, and spinal problems, with severe and congenital skin conditions. 'Over 50 of the children are already suffering from arthritis and similar conditions, although they are only now entering their 30s,' she wrote. In her study, the average age of death for veterans was just 56, two-thirds due to cancer. *Saga Magazine* noted that year that, if all surviving veterans were given a war pension with an average disability assessment of 50%, 'it would cost barely £50million a year' - or 5% of the war pensions budget.

In 2002, the *Sunday Mirror* traced 350 veteran families who had reported health problems in their children at the start of the justice campaign. It discovered 115 now reported issues with 169 of their grandchildren. They included breathing problems, spina bifida, premature births and miscarriages, bone problems and high rates of epilepsy and mental disability. Notes taken at the time form a list of horrifying conditions: 'Immune breakdown, wheelchair bound'; 'deformed feet, several ops to repair bones sticking out of feet'; 'malformed eye and ear, skin flaps growing over both'; 'son 1963, insides deformed, died after 14 months'; 'giantism'; 'deformed penis'; 'concave bone deformity'; 'born with half a kidney'; 'malformed vocal chords'; 'three daughters bone diseases, two daughters, anaemia'; 'six toes on one foot'; 'deformed – deceased 1989'; 'extra thumb, lumps taken from cheeks'; 'three nipples'; 'two thumbs on one hand'; 'heart surgery 16 days old'; 'born with no anus'.

The results of the newspaper survey were assessed by respected statistician John Urquhart. He found the rate of leukaemia was 6 times the national average, Down's syndrome was 8 times more common, and the total number of serious birth defects was triple what it should have been. And these were conservative estimates, as those families who could not be contacted were assumed to be healthy. All these conditions are things that can occur naturally, but rarely. For example, it is known there is a higher risk in older mothers of the genetic mutation that leads to Down's syndrome. Yet the ages of mothers the *Mirror* found were just 23, 24, 26, 27, 31, 34 and 37.

The BNTVA encouraged its members to apply for war pensions, even arranging a mass-submission designed to overwhelm the system. Unfortunately the process meant each individual claimant had to provide evidence of injury - easily done if you have lost a leg in Afghanistan, harder if you were irradiated and someone else was taking the notes. As almost no veteran had been responsible for taking any such records, they had to ask the MoD for the proof they wanted to use against it; something that in almost every case was not forthcoming. In 2012, according to an FOI about the minutes of a meeting held in Downing Street between BNTVA representatives and the then-Minister for Veterans, only 30 out of 1,800 test veteran war pensions had been successful, 25 of them on appeal - a success rate of just 1.6%.

In any event, most men were denied a pension before they started, because they had no dose badge. Whenever a pension was asked for, the MoD wrote to the Atomic Weapons Establishment - which had dropped the 'research' from its name in 1987 - asking if it had any records. The NRPB reports found that, of the 22,000 servicemen sent to the bomb tests over 15 years, just 23% overall had ever had a dose badge. Of those, 64% were claimed, unreliably, to have recorded zero radiation.

A New Zealand Defence Force study into Operation Grapple was declassified in 1996, and reported that when the badges were collected and sent for processing on HMS *Narvik*, most were not developed due to problems storing the chemicals. Other documents showed they did not work in extremes of heat or cold, and radiation meters used in the aircraft, like the one tested in Eric Denson's experiment, also malfunctioned. While 96% of veterans at Hurricane were given a film badge, it was as low as 2% at some of the Grapple tests. The MoD argued this was because earlier bombs had produced such low doses, that future servicemen did not need them; veterans say it is because the later, bigger, tests were conducted in haste, and top brass no longer gave a damn.

Six times at Maralinga, and twice at Christmas Island, they had been subjected to clean-ups, pronounced clean at a subsequent survey, only for more contamination to be found. In 1998 the AWE sent another team to make a 'liability assessment' at Christmas Island, which uncovered 21 luminous dials covered in radium paint. Five years later, an environmental firm found another 47 dials inside rotting military vehicles. In 2005 a company called SEC was given the contract to remove any military waste, including asbestos, bitumen, and corroded metal, with help from islanders. They found far more than was expected, including another 90 radium dials, which had rotted and caused 'localised ground contamination' which had to be dug out, bagged, and put in steel drums. There was 3,471 tons of scrap metal, 2,000 tons of concrete which was dumped at sea for 'shoreline protection', and the islanders asked for the bitumen - which had previously been found to be contaminated - to be processed for reuse in road mending.[21]

The islanders demanded the rest of the low-level radioactive waste be permanently removed. Fiji was suggested, but it was ruled politically impossible - the commander of their armed forces was the son of a Fijian serviceman who had been seconded to the tests. Within a few years, he would lead a coup and become Prime Minister. The British realised they'd have to take the waste themselves, but there were snags. Egyptian authorities refused to let them take it through the Suez Canal, so it had to be sent across both the Pacific and Atlantic oceans, instead. The UK bans the import of asbestos, but had to authorise 383 tons of its own being returned. Then the shipping firm booked for the job found out what the cargo was, and refused to carry it. In the end the vessel took everything except the most radioactive waste, which had to be flown by air. It was destined for the village of Drigg in Cumbria, where the Nuclear Decommissioning Authority runs a facility dealing with waste from hospitals, universities, the MoD, and nuclear power stations. But in the end it was sent to the Port Clarence landfill on the north bank of the River Tees, near Middlesborough.

In June 1990, while UK High Commissioner Brian Barder was being taken on a tour of Maralinga to see how clean it now was, scientists discovered part

Rocket launchers left behind after the atomic testing at Maralinga. Hundreds of rockets were fired through each cloud immediately after detonation to take samples in the very first seconds of the chain reaction. (Courtesy of Daniel Everitt-Lock, True Perspective Films)

of a uranium bomb core lying on the ground. 'Visibly-identifiable' plutonium-contaminated debris, as well as plutonium oxides which to the naked eye look like any other grain of sand, and contaminated soil was found yet again. In places, the red dust had blown away to reveal once-molten plutonium beneath: the heavy price paid for the 'burning jets' fired into the sky three decades earlier. The Australians made a claim and in 1993 the UK coughed up £20m of the £60m it would finally cost to rehabilitate the Outback it had so grievously irradiated. The Australians gave about £6.4m of it to the Tjarutja people, in full and final settlement of their claims.

Over the next five years, the Australian Radiation Protection and Nuclear Safety Agency (ARPANSA) buried a further 200,000 cubic metres of topsoil, and found huge amounts of plutonium-239. A committee was set up with local politicians, the British government, and indigenous people, and this time the aim was to make it safe enough for them to remain. The 'acceptable risk' was set at 5 mSv,[22] – about three times greater than the background dose for the average Australian – but consideration was given for the first time to the fact that traditional lifestyles were particularly likely to increase the risk of inhaling contamination, or get it in a wound. A long clean-up began, scooping up soil from the Taranaki and Minor Trials areas, and burying it in a trench with 5 metres of clean rock and soil above. Sensors were strapped to the bull-bars of pick-up trucks which drove across the range in grid patterns, using GPS to log and record all radioactivity. At the Vixen B sites they found visible fragments, including contaminated metal, plastic and wire. There was no way to know how much plutonium was in the 22 pits the British had earlier dug, but it was three times more than expected. The solution was to heat the contents until it melted, then allow it to cool and turn into glass. When something flammable in one pit exploded, the Australians decided to abandon the expensive heating, and just exhume and rebury the material in shallow trenches. There were claims even this clean-up was a cut-price job, and bedevilled by racism. In 2002 nuclear engineer Alan Parkinson, who was employed as an adviser to the Aboriginal people affected, told the *Australian Broadcasting Corporation*: 'What was done at Maralinga was a cheap and nasty solution that wouldn't be adopted on white-fellas land.'[23]

ARPANSA said this last clean-up had reduced contamination by a factor of five, and that unless someone were to start digging they'd be unlikely to inhale anything approaching the normal annual background radiation dose for Australia. Its final report concludes: 'Lifestyle changes could also markedly affect dose estimates. If in time the Maralinga Tjarutja were to move towards a more European lifestyle, with extensive areas being covered by concrete, tarmac, buildings and lawns, and living in Western-style houses in suburban settings, then the dust levels and hence doses are expected to be much lower.'[24] Garden like an Englishman, and you'll be fine.

War pensions are not approved on the basis of precedent. Instead each veteran must re-prove the case, using records they have no access to, from monitoring that was not thorough, about an incident that took place a lifetime ago. In 2003 a third NRPB update found there was still no good reason for the test veterans' continuing concerns, which effectively stymied any claims.

Roy Vincent, a driver in the Australian army, pictured in 1961 in the Bedford truck he used to transport Radiation Detection Unit officers to the 'hot areas' of Maralinga. Dust kicked up by other vehicles spread radioactive particles on the breeze. (Courtesy of Roy Vincent)

Roy Vincent pictured on a recent return to Maralinga, next to the warning sign telling visitors the area still contains radioactive artefacts. (Courtesy of Roy Vincent)

Three years later there was a small breakthrough. Roy Prescott had served as one of 500 Royal Engineers at Operation Dominic for 11 weeks in 1962, and thought nothing of it until the 1990s when he began to hear of comrades falling ill. Non-smoker Roy, from Burton-on-Trent, Staffordshire, later developed small cell lung cancer, a rare bone disease, and had three tumours removed from his head and neck.

He was turned down for a war pension by the MoD, which claimed there was 'insufficient evidence'. He then applied to the Americans, and was awarded $75,000 because they recognised his cancer as radiogenic. He said at the time: 'I am a casualty of the Cold War and, whilst I am pleased that I am receiving compensation and recognition from the US government, it really galls me lying here, a critically ill man, that the British government continues to fail in its duty of care towards me and thousands of other nuclear test veterans by denying that we were exposed to radiation during service.'

He died a few weeks later, aged 66. Three months after that, an appeal was heard about his UK war pension. His son Craig spoke on his behalf, and the tribunal agreed that Roy's exposure to radiation had probably led to his illness. Craig said: 'It's too late for him now, but as a result of this my mum will be secure for the rest of her life, and perhaps other veterans can win the same... My dad died very angry about the fact that he and all his mates had suffered but never got any recognition, and he fought for years to win compensation for all of them.' An MoD spokesman claimed that the American compensation scheme was less stringent than the British one which requires 'those claiming compensation to show a reasonable link between their service and their illness'.

Four years earlier, Pat Spackman had applied to the MoD for a war widow's pension after her husband Derek, an RAF navigator, was ordered to fly sampling missions through the clouds of six US Castle tests in 1954 as part of Operation Bagpipes. Derek spent up to two hours in the clouds, and later died from throat cancer just five months after diagnosis. Pat was refused a pension in the UK, and lost again on appeal in 2004. But in 2010 she won a payout from the Americans. She said: 'This makes our own government's attitude to Derek, and all the other nuclear veterans, even more shameful.' Her stepson David Spackman added: 'I am saddened and angered that the UK government, which my father served without question for so many years, has refused to accept that flying through an atomic mushroom cloud may even possibly have contributed to his throat cancer.'

In 2004 a group of 1,011 veterans launched a civil suit against the MoD for negligence, and in 2007 they thought they found their silver bullet – research from New Zealand that showed test veterans had genetic damage. Finally, this was proof that the radiation had got under their skin. The study was conducted by geneticist Professor Al Rowland of Massey University on 50 of the 551 Kiwi sailors who took part in Grapple. The aim was to examine the rate of translocations in veterans' genetic code; in plain English, whether their DNA had been scrambled.

DNA is a chain of molecules in each cell that provides a set of instructions for how it should work. As you age, and depending on lifestyle, bits of the chain break. Sometimes the broken bits reattach - or 'translocate' - elsewhere on the chain, and as a result the cell is given different instructions. It is a normal part of the ageing process, but it can also be influenced by factors like smoking. And as up to 75% of the human genome is thought to be a defunct hangover from our evolutionary past, some changes make very little difference, while others can harm. Rowland's team found nuclear test veterans showed a rate of genetic damage three times that of the ordinary population.

The nature of radiation, and the immense variability of genetics, means not every exposed person experiences the same effects. It can best be demonstrated by the air crews, who were always first in harm's way. After the first blast in Buffalo, a sniff plane from 76 Squadron entered the cloud at 8pm, with another mission around 1am and a third sweep at 3am. On the third plane was 24-year-old flight lieutenant Barry Newton. He flew three more missions during Buffalo, and a year later volunteered for another three studying the clouds of the first Grapple tests. He was later aide de camp to the Queen, and retired as Air Vice-Marshal. His obituary in *The Times* in 2020 records that 'he never suffered ill-health as a result of the nuclear tests, even though he was probably one of the most irradiated men in the country. As his daughter said, there were people who thought he "glowed in the dark".'

The same article records the very different impact of the tests on another 76 Squadron member, Christopher Donne, who took part in the Halliard test in 1958. He stayed in the cloud for 12 minutes. He later recalled: 'I remember seeing this yellowy-brown thing ahead of me, stretching out almost as far as I could see, and I remember turning the aircraft and getting it straight and level and just scrambling up those last few feet and then approaching the cloud... And then we hit it, and I can remember my navigator saying "Bloody hell! Let's get out of here!" But of course, we couldn't because there was no way I could turn the aircraft - the turbulence was causing me to concentrate very hard on flying it at all at that height. I can remember sort of glancing out of the side of my eyes to look at the instruments - the needles pressed very firmly up against the stops... showing the very high levels of radiation, which were very much higher than we'd anticipated. I can remember the health physicist muttering in his beard something about it being very much hotter than he'd thought.' On landing, the crew were decontaminated and Donne was told to return to the UK after exceeding his dose limits. He had operations to remove skin cancers in the 1990s, and went on to develop further skin, bone and bladder cancers, and to survive quintuple heart bypass surgery. He was diagnosed with prostate cancer in 2003, and was nursed at home by wife Christine until his death 12 years later, in 2015. She, like so many others, was refused a war widow's pension.

Rowland's study compared the DNA of survivors with that of a matched group of 'controls', who were of equivalent age and health, also serving in the navy, and with the same proportions of smokers and non-smokers. He said later: 'My study showed that, even taking into account confounding factors that can also damage DNA like smoking, radiotherapy and exposure to solvents, the veterans had damage similar to that found in clean-up personnel at Chernobyl. We struggled to find 50 veterans to study because we had such strict guidelines about who could be included. There is no trace within the DNA to say it was the bomb tests that did it, but considering how other factors had all been excluded I struggle to think of another factor that could result in such drastic effects. Genetic damage is observable and most likely to be due to exposure to radiation.'

He added: 'The science isn't there yet to say for definite these defects are due to radiation, but when there's such anecdotal evidence it would not be surprising to me if it was found in the end to be down to that initial exposure. There are new techniques which could unfold the story.' At first the MoD refused to comment on the research, and after it was accepted by peer review still insisted it was not enough proof. It claimed the study 'left open the question of whether all of the dose was delivered around the time of the nuclear weapons tests'. It did not indicate what other source of radiation could have delivered the same damage, to the same people, before or after the weapons tests. In 2012, the Queen gave Prof Rowland the New Zealand equivalent of the OBE for genetic research.

In the same year as Rowland's study was published, Professor Chris Busby at the University of Liverpool published the results of another postal survey conducted with the BNTVA. Its members had been asked to list the health conditions of themselves and their families, with Dr Busby analysing the results. He discovered the children of veterans reported 10 times the normal number of congenital birth defects, that wives had three times the usual rate of miscarriage, their children were five times more likely to die as infants, and their babies were three times more likely to be stillborn. Veterans' grandchildren were eight times more likely to have birth defects, and twice as likely to get childhood cancer. But again the MoD claimed the research was flawed. Not only were these self-selecting individuals who had joined a campaigning group because they were ill, but Busby was a former Green Party candidate and anti-nuclear campaigner. The study was peer-reviewed, however, in 2014, which gave it scientific approval, with some caveats.

The peer review stated: 'Whilst caution must be exercised due to structural problems inherent in this study, we conclude that the veterans' offspring qualitatively exhibit a prevalence of congenital conditions significantly greater than that of controls and also that of the general population in England. The effect remains highly statistically significant even assuming a high selection bias in the responses and credibility is strengthened by the high rates of miscarriage reported by the veterans compared with controls, a result which could hardly have been due to selection effects.' The MoD did not alter its view.

In 2009, the veterans' case finally came to court, but their evidence was not heard in full. Instead, the MoD argued they had no right to a trial at all.

The survivors brought their case on the legal basis of causation - that their health problems were caused by exposure to radiation. But the MoD and its legal team fought the case on the grounds of time limits. Under health and safety legislation, a victim of employer negligence has to bring a claim within three years of 'effective knowledge of injury'. One of the lawyers acting for the veterans explained to the *Daily Mail*: 'A limitation defence means that instead of getting to the meat and bones of whether the MoD is liable for the illnesses and injuries suffered by these men, a case first has to be presented to the court and argued that the action brought against the MoD was within the right timescale. If the argument fails, the case ends at that point without the court ever hearing the real issues and evidence in the case.'

Ten claimants were selected as examples, and questioned forensically about when they first made the link between the bomb tests and illness. They included Ken McGinley, Wendy Brothers and Archie Hart, as well as Christopher Noone, an RAF crewman who saw Grapple X and afterwards flew back to Britain with radioactive samples behind his seat. He developed 100 blisters on his back, lost all his teeth, and over time became depressed as his back never healed, leaving shirts soaked with blood and pus. Also selected was Roy Ayres, a RAF regular who had developed prostate and bone cancer after serving at Grapple Z, and kept weeing blood.

Another test case was Rosie Clark, a widow bringing a claim on behalf of her husband Mick who had died 17 years earlier, aged only 54, with cancer of the lung, cancer of the lymph glands, and a rare type of bone cancer. Taxi driver Mick had been a Royal Engineer and on National Service during Grapple. On his return he married Rose, and had three daughters. In the 1980s he had a mysterious rash all over his groin and upper legs that took months to clear up, and in 1990 began having flu-like symptoms. A few months later he collapsed and an MRI scan revealed his body was riddled with cancer, with secondaries in his back, ribs and bones. Before he died Mick wrote an account of his experiences. It says: 'One vivid recollection after one detonation was a fierce thunderstorm which was followed by torrential rain… it rained so heavily that we were floating our cigarettes around to each other inside our tents. At no time ever was I examined or asked whether I felt all right, nor was I at any time given a badge to wear with which the fallout was supposed to be measured. After one such detonation I came out in a skin rash which took six weeks to clear… during the five tests I witnessed, I was only given clothing to wear on the last two and this consisted of nylon trousers, top, gloves and hat.'

Rosie joined three other widows, Jean Sinfield, Ethel Ogden, and Evelyn Dickson, bringing claims for their husbands for leukaemia, brain tumour and colon cancer, and heart disease and diabetes, respectively. Finally, there was Pita Rokaratu, a giant of a man who had been a Fijian sailor and acted as a stevedore, loading and unloading ships at Christmas Island. He recalled seeing 'the seawater bubbling' after the blasts, and being dressed only in shirt, trousers, boots and a cap. He went on to lose his hair by the fistful, have stomach problems, and saw the same lipomas as Archie Hart had appear all over his body. He developed aplastic anaemia and arthritis.

The hearings lasted for weeks. McGinley had a torrid time in the witness

Pita Rokoratu was a Fijian sailor ordered to take part in Britain's nuclear tests as part of the Commonwealth. He worked as a stevedore, unloading ships, but despite ill health he travelled to London to appear in person at the High Court. (Courtesy of Pita Rokoratu)

box, with the MoD barrister insisting that, as he had formed the BNTVA in 1983, he had left his claim 20 years too late. Every witness who had been quoted in a newspaper report was claimed to have 'known' the source of their injury was radiation, even though the official MoD line was still that there had been none. Finally, the MoD asked the judge to throw the claim out. All 10, it argued, should have sued earlier. The Rosenblatt team said the veterans could not have done so without the Rowland study, which had made a no-win, no-fee arrangement possible, without which such court action would be beyond financial reach. Six months later, Foskett made his decision - and ruled the veterans were 'entitled to their day in court'. The MoD was facing a full trial, and this time it would have no choice but to defend the charges of negligence in exposing servicemen to radiation.[25]

The veterans were jubilant, and shared glasses of champagne on the steps of the High Court. But the MoD disagreed. It appealed the decision, and won. The veterans counter-appealed, to the highest authority in the land - the Supreme Court.

In March 2012 it ruled - by the narrowest possible majority of four judges to three - that 9 out of 10 veterans' cases were brought too late. One judge, whose view swayed the majority against the veterans, wrote that they had 'no prospect of success' in proving their case. Lawyers for the veterans felt such an assertion to be extraordinary - how could anyone say whether the veterans could prove the case, when the case had not been heard? While most of the veterans were infuriated, and media reported on their 'failure', few realised they were denied the right to present evidence, and a Supreme Court judge had nevertheless ruled it hopeless.

After reading out the judgement, one of the justices, Lord Wilson, said: 'Putting aside the law for one moment, all seven members of the court would wish to record their personal sympathy for the veterans... It must be bad enough for the nine veterans, and the other claimants, to learn that they have lost this final round, but to learn that they have lost by the narrowest possible margin must make it even worse.'

Speaking outside the court to the BBC, Trevor Butler, of Hull, told how serving with the Royal Engineers during Operation Grapple had left him with a crumbling spine, kidney problems, and skin complaints. He said: 'I'm absolutely devastated. After all the years of pain and anguish how can they do it to us?... What we have suffered is completely irrelevant to the MoD. They have just brushed us aside on a legal technicality.' Rosenblatt had racked up a £17m bill, through years of work wasted gathering evidence of causation, and more years building the time limitation defence. And by the time the Supreme Court had ruled, five of the test cases had died.

The veterans' lawyer Neil Sampson said: 'The approach that this government takes is to waste resources on fighting veterans, rather than co-operating with them. There are some things in life that are wrong. The approach of the government to this issue is one of those things.' The MoD put out a statement, referring to that single judge's bizarre assertion about the evidence, saying: 'The Supreme Court described the claims as having no reasonable prospect of success and that they were doomed to fail.' Had a judge pronounced the veterans were bound to win without having seen any of the evidence, it is unlikely the MoD would have accepted it.

The veterans had been denied their day in court, and just like the sticky ash that fell at Wallatinna, it left a bitter taste in the mouth. The weapons tests became a scab that was constantly picked - by the families, by the *Mirror*, and by those who couldn't understand why the case failed. In 2013, Labour MP Kevan Jones - who was Veterans Minister at the time of the hearings told Parliament: 'A generous settlement proposal was put to the lawyers - I got the Treasury to agree to it - but it was rejected. That was an opportunity missed for veterans to get some compensation.' He was wrong. In fact, all sides have confirmed they were never given an offer to reject.

When he was minister in 2010, after the High Court had authorised a trial, and before the appeals were heard, it looked like the MoD might have to find more than £20million in costs alone, plus compensation. It was politically obvious that it would be cheaper to settle.

The British legal system has two tiers. Someone with a complaint may hire a solicitor to act for them, and if the dispute gets into a higher court or needs specialist help, that solicitor will instruct a more senior lawyer called a barrister, also known as counsel, to act on their behalf. Ethical rules make it clear that both sets of lawyers must keep their clients informed, offer them advice, but ultimately must follow their instructions.

Jones, as minister, ordered his officials to make an offer. This was confirmed by the MoD directorate of judicial engagement policy in a Freedom of Information request in 2019. It said: 'Mr Jones authorised a range of figures within which, if settlement could be reached, an offer could be made. Discussions between legal representatives (counsel and solicitors) for the nuclear test veterans and the MoD in early 2010 revealed that no offer the MoD legal team were capable of making would be recommended to the veterans for acceptance, indeed they would be told to reject it. Therefore, no offer was made.'

If this statement is correct, it amounts to a series of astonishing allegations. Put simply, the MoD claims a meeting took place between lawyers from both sides, an offer was not made but was rejected anyway, and then neither side told their clients what had happened - not the veterans, or the minister. To be true, it needs a dozen lawyers to be equally unethical, and for a legal firm that had run up a £17m bill to refuse to even talk about letting someone else pay it.

The FOI says the veterans would be 'told' whether to take the money or not. Yet that is strictly against the rules. Offers should be communicated to clients, recommendations can be made about whether or how to accept it, but the decision is ultimately that of the customer, not the lawyer. The only way the MoD statement can be true is if two entire legal teams ignored professional guidelines, managed without mentioning a figure to decide it wasn't enough, and failed to tell the minister what had happened to his 'offer'.

The official record shows that the veterans' lawyers asked the High Court to order mediation to settle the dispute, which would not have happened if an offer had been made. They also wrote repeatedly to the MoD requesting a meeting, and for details of the 'alleged offer'. Each was refused.

What actually happened is hard to unravel. One lawyer involved with the case claimed the government's barrister invited his opposite number to a meeting, with a highly-unorthodox pre-condition: he would not be able to tell his clients about the meeting, or the offer. If this is true, it is likely to have been an order from within government, with the barrister rightly following the instructions of his client. But for it to have been agreed would have made all talk of an offer moot, because the veterans' barrister would be unable to pass it on to the only people who could decide whether to accept it. To demand the barrister keep the whole thing under his hat, and invite him to accept or reject the offer on their behalf, was not just unusual or unprofessional. It was impossible for anyone who wants to still be a lawyer the next day. The veterans' barrister refused the pre-condition, and nothing more was known about it until Jones mentioned it years later in Parliament.

What the politician had instructed his civil servants to do somehow never materialised. Somewhere in the chain of command, an impossible demand was inserted, which derailed the agreed policy of a democratically-elected government. The instructions of a Minister of the Crown were rendered null, by someone who has never been identified.

Blocking the payout bought time for the MoD to appeal, and get the case thrown out. The legal bills were so vast that the veterans' lawyers were left to lick their wounds, and other firms were dissuaded from entering the fray. After nearly a decade of legal wrangling, those who survived had to battle on alone.

Shelly Grigg's father Roy was at Christmas Island for a year during clean-up operations. He swept up sand, ate the fish, and swam in the lagoons. He later became a priest, and died of aggressive bone cancer in 2001, aged 66.

Shelly was diagnosed with rheumatoid arthritis at 15. At 22 she had a massive growth on her knee removed. She said: 'In my 30s I got painful stomach lumps called lipomas. It's Dercum's disease and is very rare. In a test veteran families' support group I found six people whose fathers were at nuclear tests or worked in the nuclear industry with the same illness. My spine is crumbling. I've had discs in my neck replaced with metal.' Shelly has been told she has the body of an 80-year-old and has to take 14 pills each day. Her only comfort is she never had children. 'I have no way of knowing what I would have given birth to,' she said. 'We are the fallout of those bombs. The children and grandchildren of the men who were there are in chronic pain, traumatised and ignored. We need medical help and support, but we're left to suffer because of something the government did to our dads and now doesn't want to face up to. But we have to face it every day.'

Today, the cancer and birth defect statistics for Christmas Island are subsumed into the overall figures for the Kiribati Republic. But according to the Kiritimati Association of Atomic Cancer Patients, 189 families reported illness after the tests, including cancer, blindness, deafness, birth defects and high infant mortality. In *Grappling With The Bomb*, Sui Kiritome recounts being invited onto the deck of a warship with her schoolteacher husband to watch Grapple Y. She said: 'I felt something like a light shower falling on me. I thought it was rain... I felt wetness on my head, my face and skin.' In the days that followed, what had seemed a kindness from the ship's captain to partake in a spectacle took on a darker tinge. 'Every time when I combed my hair, I was losing strands of my hair and something like burns developed on my face, scalp and parts of my shoulder,' she said. 'The mark remains on my face til today.' Sui was pregnant at the time of the bomb, and says her daughter was born bleeding from eyes, nose, and ears.[26]

Makurita Baaro started school on Christmas Island in 1963, and later became Kiribati's ambassador to the UN. 'One of my classmates had no teeth. She never had teeth. Another boy in the same class had patchy white and brown skin,' he said. 'In Kiribati, no studies have been done on the effects of these nuclear tests on our people... our communities still suffer the long-term impacts of the tests, experiencing higher rates of cancer, particularly thyroid cancer, due to exposure to radiation. In some places, the environment and food sources remain highly contaminated, although studies done by those who conducted the tests have cleared these areas and indicated they are safe.'[27] Similar stories are told across the Pacific, where France and America conducted tests, near Soviet and Chinese test sites in Asia, and among those who live in the Outback.

Christmas Island has little to show for the crucial role it played in forging modern Britain. Its people give birth in a Portacabin, have a poor diet, and as a result of climate change are seeing their island become barren. Their main source of income is tourism from deep sea fishing. Academics have argued that respect, at the very least, is due to the people of Christmas Island, and have called for reparations and medical help. The British government responded by telling Parliament in late 2023: 'The UK has reviewed residual contamination relating to British nuclear tests on Kiribati and concluded that any required remediation had been undertaken.'

It was not until 2009 that contamination was considered low enough for 3,000 sq km of the Australian desert to be returned to its original custodians. A further 412 sq km were opened for access but ruled out for "permanent" habitation. Today, a boundary track is marked every 50 metres with signs saying 'kaka palya' (meat good) and 'ngura wiya' (no camping). Continuing low-level plutonium contamination makes it risky to sleep or sit on the ground, and campfires are best avoided.

In the 1950s, sacred objects were removed to the missions as well, and native patrol officers recorded that Maralinga no longer had any draw for indigenous people. But in truth the spiritual connections, to places of birth, death and dreaming, remained intact. These were the places to which ancestors had travelled for ceremonies, and where their descendants still wish to go for spiritual events.

Many of those who were evicted never went back. Most now live at a settlement called Oak Valley, 75 miles north west. Jeremy Lebois, chairman of the Maralinga Tjarutja, said: 'For my grandparents, that land they saw it as "mamu", an evil place. It was a land they could not return to. The people today, we understand about radiation and atomic testing and wouldn't return to live there, but we would visit it.'[28] The University of Sydney found they reported - just as veterans do - sickness, high rates of miscarriage, and birth defects, but without medical or hospital records have been unable to prove a link with radiation to the satisfaction of the authorities. In 2011 the Taranaki clean-up trench was reported to have eroded, along with 19 out of 85 of the pits. The Aboriginal caretakers were blamed for not taking better care of it.

There is one more potential victim of the nuclear tests who deserves to be noted. William Penney was given an OBE, knighted, then made a life peer, for service to his country. He died of liver cancer in the year of the last British nuclear test, aged 81. His death certificate says the main cause of death was 'carcinoma of the liver' but it was thought to be a secondary cancer, caused by a primary tumour elsewhere. No main tumour was ever found. Professor Chris Rhodes, a radiation specialist who worked with victims of the Chernobyl disaster, told the *Mirror* in 2009: 'If they can't find the tumour, how did they know the liver cancer wasn't the primary cancer? It sounds as though somebody is trying to wriggle out of responsibility.' The government's studies have found test veterans are almost twice as likely to get liver cancer, and the US government pays compensation to test veterans who contract it. Experts agree that the radiation he spent so long working with put the chief architect of the tests in harm's way. GP Chris Steele said: 'If someone has been playing with uranium for a large part of their adult life, you're at very high risk of just this type of condition.'

The same article reported that Penney's son Martin, a retired headteacher, said the tests' legacy of illness was 'a tragedy for many people'. He added: 'We have no idea what caused my father's cancer. Lots of people have complications at his age.' The only time Penney commented on the tests was at the Royal Commission. He was distressed by the hearings, feeling he'd been made a scapegoat. He told the inquiry that the public were 'kept in the dark' about the size of the bombs, and that fallout forecasts turned out to be 'drastically off-course'.[29]

He added: 'I thought we were going to have a nuclear war where the only hope I saw was that there should be a balance between East and West – but what I really wanted was to be a professor.' Penney confirmed that 'radiation hazards will be present' before, during and after the explosions, and that he took advice from the Medical Research Council on the 'safety regulations' as he wanted no person to be knowingly exposed. The mathematician, probably reasonably, relied on the medical men to set the tolerance limits, while he focused on the physics. His son said: 'My father didn't want to do the job at all – he was just doing what the Prime Minister of the day, Clement Attlee, asked him to do.'

Yami Lester died in 2017, having won recognition for the suffering of his people. Australian Prime Minister Malcolm Turnbull called Yami 'one of the most significant Aboriginal leaders our country has ever known'. In 2021, 65 years after it was first used for radiation experiments and after seven clean-ups, a team of researchers from Monash University found Maralinga was still contaminated. They analysed soil from the Taranaki blast collected in 1984, and found that fallout particles once considered stable and inert were in fact volatile, and likely to break down in the harsh desert environment. Thanks to the rapid heating and cooling particles had undergone in the weapons tests, they were more likely to be inhaled, taken up by plants, and had already entered the groundwater. It meant that those who had returned to their homelands, and who considered they had a cultural responsibility to help the land heal by using it to hunt, gather and cook native foods, are more, not less, likely to be exposed.[30] Today the Tjarutja offer visits to the test sites for tourists. For A$300 a day, a local will provide a living history lesson about one of the darkest chapters in the Australian story. Protective clothing is not necessary, but the website states: 'Put in perspective, for the few hours in the Maralinga forward area you might receive as much radiation as a couple of days elsewhere.'

The area's uranium mines are still operating, and export their product worldwide for use in nuclear technology - power stations, medical equipment and weaponry. Throughout the 1990s there were hard-fought proposals to establish nuclear waste dumps at Maralinga, and it is easy to see how, to the authorities, it was the best place for it, while to its traditional owners there was quite enough already.

Perhaps the last word on the Australian tests should go to Dr Keith Lokan, director of the Australian Radiation Laboratory, who gave evidence to the Royal Commission and was asked how long Maralinga should remain under scientific supervision.

'A million years,' he replied.

EXPOSURE

FLEET STREET LEGEND Richard Stott is the only person to have edited two British national newspapers, twice. He led huge teams at both the *Daily Mirror* and its sister paper the *Sunday People,* two tabloid 'red-tops' selling to millions, producing scoops, exclusives, and investigations with a left-wing slant. Both have a proud record of long-term campaigns on behalf of those who are unheard or victims of an injustice, whether it was families affected by the Hillsborough football disaster, the striking miners of Orgreave, or children in need of organ transplants. Although today they are battling a drop-off in print sales as a result of the internet, and massively reduced budgets and staffing levels, the campaigning continues, forcing politicians to take note and, often, to change the law.

Stotty felt he worked for the readers, not the owner, of his newspaper. He said: 'I believed in what the *Mirror* stood for - social justice, decent and honest standards in public life, and the right for people with small voices to be heard loud and clear.'[1] This instinct for defiance infected others, and led to a fight for justice on the nuclear veterans' behalf that has become the longest-running campaign in British newspaper history, outlasting both his editorship, and him.

In March 1984, Stotty ran a front page in the *People* headlined 'Our A-bomb Guinea Pigs!', detailing the declassification of that 1953 memo to the Chiefs of Staff saying the military

Legendary Fleet Street editor Richard Stott started a campaign for the nuclear veterans in 1984 which continues to this day - and is now the longest-running newspaper campaign, about the longest-running scandal, in British history. (Courtesy of the Stott family)

'must discover the detailed effects of various types of explosion on equipment, stores and men, with and without various types of protection'. It led to hundreds of calls to the newsroom, as well as a bulging mail bag of letters to the editor. Stotty realised this wasn't just any old story. Six months later, he launched a campaign called 'Atom Bomb Kids', with another front page telling five million readers the results of an investigation into the families' birth defects. 'Fifty seven children of British servicemen exposed to nuclear bomb tests have died, been hideously deformed, disabled, or have other crippling illnesses,' the paper reported. 'But scientists are deeply divided on whether or not people exposed to radiation can transmit disability and disease to their children. Disturbing evidence that such a link MAY WELL exist emerges from our investigation.'

The paper spoke to 32 veterans, including one who served at Maralinga and whose son died aged three, described as 'so terribly deformed that doctors did not want his parents to see him', with missing fingers and toes, and a face 'like an old man's'. The wife of another was sterilised after having two deformed children. A Grapple survivor told how his son was born with a bone-crumbling disease which killed him aged 12, and spent his life in pain but was unable to produce tears because of another disability. 'He sort of cried inside and rocked back and forth when the pain got too bad,' his dad said.

The paper taglined it 'a story to make your heart weep', and it was. But it was unbearable to read, with the consequence that people were outraged about something they'd rather not hear about. Pictures of dead and deformed children, grieving mums, and sickening servicemen were of vital importance, and treated sensitively for the time, but it was not the stuff of which extra newspaper sales were made, as the media was swiftly coming to learn.

Others were also covering the story and in an industry battling the loss of readers to television, it was difficult for editors to insist it go on the front page, where anything too horrifying would drive people away. As MoD denials continued and veterans were unable to make headway, newspapers lost interest. Stotty, alone, did not. Under his editorship the papers continued banging the drum for years. By 1997 when Blair was swept to power and turned his back on the veterans, Stotty's impact was such that, even though he was no longer editor, the *People* still ran features on 'the A-tests scandal'. At the end of that year, its leader column thundered: 'Today on these pages we reveal the heartbreaking story of a new generation of children blighted by the atomic tests witnessed by their grandfathers 40 years ago. Born with terrible deformities, these children must be cared for. The *People* has campaigned for justice for these veterans and their families. Now a third generation has inherited the Bomb's evil legacy and needs our help. It is a campaign this newspaper has yet to win, but we are determined to try.'[2]

Yet without public or political impetus, the campaign faded to occasional updates. In 2002, by then the grandee columnist of the *Sunday Mirror*, Stotty rang his editor – Tina Weaver, a former protégée – to ask what she was doing to mark the anniversary. 'Of what?' she asked, innocently. 'Operation Hurricane!' he blistered back at her, furious that anyone could have forgotten. Journalists were set to work,

Sunday People

ANOTHER PEOPLE EXCLUSIVE

MARCH 11, 1984 No. 5325 25p ★ L

GO GET 'EM GIRLS!

A Prince each for Katie and Romy.. but who do YOU prefer? CENTRE PAGES

EXCLUSIVE STRANGE TRUTH OF 'INVENTED' CROSSROADS ROMANCE • SEE PAGE 19

After 30 years the secret is out

OUR A-BOMB GUINEA PIGS!

Soldiers WERE exposed in 'killer' tests

By CHRIS BUCKLAND
Political Editor

THOUSANDS of British soldiers were deliberately exposed to deadly radiation during atomic bomb tests in the Fifties, it was revealed yesterday.

Some of the human guinea pigs were sent into the danger area without protective clothing according to documents just released from the secret list.

The astonishing revelation is made in a memo to the service chiefs from the Defence Research Policy Committee.

The memo, dated May 20, 1953, has been discovered among documents released under a rule which de-classifies secret documents after thirty years.

It will come as damning evidence against the Defence Department, which has always refused to accept responsibility in the face of servicemen's claims for compensation.

No one has ever been able to estimate how many servicemen became ill after taking part in the tests, which took place at Maralinga in Australia and Christ-

Continued on Page Four

TOO CLOSE? An A-test.

Jackpot PRIZES GALORE On Page 32

BINGO WIN £5,000 Page 21

The *Sunday People*'s front page which launched the fight for recognition: 'Our A-Bomb Guinea Pigs!' It quotes a 1953 memo showing top brass wanted to test the effects of radiation on servicemen and equipment. (Reproduced with the permission of Reach plc)

and that October feature writer Alan Rimmer, who had worked on some of the early stories in the 1980s, bylined another front page: 'CURSED.' Above an image of a thermonuclear cloud, the strapline read: 'Britain's A-bomb tests blighted the lives of thousands of soldiers. 50 years on, a legacy of death and disease is being passed to their children and grandchildren.' Over six pages inside ran 'an investigation that shames the government', in which Rimmer tracked down hundreds of those he'd interviewed decades before to find grandchildren were now suffering too.

Stotty wrote an excoriating column, full of fury and giving no quarter. He wrote: 'The scandal of the nuclear test guinea pigs is a lasting and ignominious stain on Britain and 14 successive governments, Labour and Conservative. They have not just turned their backs on the suffering of servicemen exposed to radiation, they have consistently ignored mounting evidence and condemned men, women and children of four generations to face lengthy, agonising, humiliating and poverty-stricken deaths.

'There have been too many funerals already, not only of the test veterans, but of their children and their grandchildren. How many more generations of the damned will our politicians allow to suffer before they accept the calamities of their predecessors and the consequences of their own cowardice? In a very few years there will be no nuclear test veterans left - old soldiers are fading away fast now.

'Often in terrible pain. Frequently alone. Certainly in a poverty ordained by a Government that continues to deny the nightmare repercussions of an experiment that destroyed them and those who came after them. And remember, lest we forget, that it is all done in our name. We have turned away from the generations we have condemned to walk with the ever-present fear of deformity, insanity, cancer, leukaemia and early death. And we have condemned them to do it without the cash help they so desperately need. That is how we really express our admiration for Our Boys. That is how we demonstrate our Duty of Care. Every man and woman in this country should be ashamed we tolerate such a legacy. And elect governments that bring hope of justice in opposition, only to deny it in power.'

Stotty died in 2007 of pancreatic cancer, aged 63. Weaver continued to push the campaign until she left in 2012. It returned to the *Daily Mirror,* and at time of writing in 2024, that newspaper is the last one in Fleet Street still fighting for the veterans and their families to be heard, just as Stotty always did.

And every word he wrote is still true today, except for one – it is now 29 British governments that have insisted no harm was done.

Two things happened with grim inevitability as the years passed. The first was that survivors died, with more memories lost every winter, and the second was that official documents became declassified as they hit the limits of the 30-year rule. All government records have an expiry date, after which legislation demands they should become public in the National Archives. While three decades is usually enough time for no-one to object, in special circumstances involving Royalty or state secrets, they can be held back for 50 years or more. But some records about nuclear veterans have been locked for more than a lifetime - with at least one file closed until 2092, by which point even the youngest veterans would be 153 years old.[3]

But information trickled out, both at home and abroad. In 1983 the US settled with the Marshall Islands, and later reached an agreement with Japan over the crew of the *Daigo Fukuryū Maru*. A decade later, a document emerged about medical examinations that had been hurriedly ordered on Marshallese after the unexpected fallout from Castle Bravo.[4] Led by US navy doctor Eugene Cronkite, it found 239 islanders had been exposed, and 28 American servicemen. Within days, they lost hair and showed significant skin damage, including ulcers and 'raw, weeping lesions' which the Brits already knew about, from Wally Kenyon. They were relocated to other islands, but in the next five years miscarriages and stillbirth doubled among exposed women, and children showed abnormalities. After 14 years researchers found high rates of thyroid cancer in children, and within 20 years a third of those exposed had developed tumours.

In 1988, junior defence minister Jack Ashley wrote to veterans who had asked for a judicial inquiry into the radiation hazards of the tests, telling them that the NRPB report proved there were no risks to inquire into. He said the study 'shows that the precautions taken to prevent harm to [veterans'] health were effective and that the chance of any of them having suffered harm is extremely small. We do not believe, therefore, that any purpose would be served by a judicial inquiry.'[5] He does not appear to have wondered if it could be true that British servicemen were the only life form impervious to radiation.

Tory and Labour governments pushed back against the veterans throughout the 1990s, although the issue didn't go away. A play about the tests appeared at the Edinburgh Fringe, another TV documentary was made, and the *Mirror* and *People* both carried regular updates on their news and letters pages. In 2002, the British archives released papers from a government committee tasked with working out how to survive a nuclear war. The conclusions it came to, for British civilians under attack from a foreign power, were very different to those the MoD claimed for troops living and working near the target areas of its own weapons. The Strath Report[6] said that air bursts were more dangerous than a ground blast, capable of total destruction within a radius of four miles and irreparable damage for another six. It said that a 10-megaton warhead - three times the size of Grapple Y - would produce a fallout zone covering 270 square miles, an area larger than the county of Middlesex, 'in which radiation will be so powerful that all life will be extinguished'. A third of the population would die. The defence strategists figured that a Soviet attack would go for the less-destructive, but more toxic ground burst, to maximise the amount of fallout that could cripple Britain for centuries. They may have been working to the worst-case scenario, but graphs show they considered everything within 20 miles of Ground Zero would be subject to fires, building damage, and burns, with a lethal dose of radiation just five miles from the blast. At the nuclear tests, servicemen had been sleeping and eating within such distances for months during multiple bombs; yet the official line on how safe Our Boys were did not change.

Two years later, a government committee of 12 radiation specialists pronounced the risks of plutonium inside the body may be 10 times worse than previously thought, and safety limits would need to be slashed. They said the 'margin of uncertainty' about the effects of ingested radionuclides 'could extend over at least an order of magnitude', which 'should be borne in mind by those making judgements and policy decisions on low-level internal radiation'.[7] Again, it made no difference to the test veterans.

By 2006 the campaign was in hiatus. McGinley had left the BNTVA in a fallout with Gray, Rimmer had left the *Sunday Mirror*, Stotty was ailing, and Blair was more worried about his legacy after declaring his intention that September to resign as Prime Minister. Two months later, news broke that Sir Richard Doll, a medical expert renowned for making the link between smoking and cancer, had been taking money from the chemical industry at the same time, as a consultant. He had also worked on the first NRPB report, and it was a chance to run out the old campaign again. *Sunday Mirror* news reporter Susie Boniface was tasked with finding out what the BNTVA thought, wrote it up for page 6, and thought no more about it.

A few days later the phone rang and on the line was Craig Prescott, who wanted to tell 'the test vets' newspaper' how his dad Roy had won a war pension from the Americans, but been denied one by his own government. The MoD spokesman was quoted saying war pensions were available. Boniface was a 'taxi rank reporter', there to pick up whatever came in, whether it was a celebrity interview, crime story, or hopping on a plane to a distant humanitarian disaster. She was quick but thorough, with a knee-jerk instinct for insubordination which annoyed every news editor who had to deal with her. She was previously a local defence reporter in the naval city of Plymouth, where nuclear submarines regularly berthed, and radiation had been a hot topic. Her father was once a civilian contractor on Renown-class submarines in the 1960s, and she could grasp the basic physics. She also knew the way the MoD was handling this story was different to others, with flat, unfriendly denials that took days to be approved for release, and made no sense. Despite joining the newspaper four years earlier, she had given little heed to the occasional campaign updates buried at the back of the paper. A brief marriage had ended badly in 2005, and her attention was given to having a good time, rather than work. That fateful phone call from Prescott's son came at the right time to the right person: someone looking for a cause, who knew how both radiation and the MoD worked, and had a naturally-contrary nose for news.

Prescott's story was followed by a steady stream of 'ring-ins' from veterans and families, who thought a fresh byline was a boost for the campaign. In 2007 came reports on Shirley Denson and Beryl Oxberry, who was suffering from a miner's lung disease after washing her husband's dusty Maralinga uniform, on Al Rowland's genetic research, and the '500-year curse' of Chris Busby's survey on birth defects. A Parliamentary inquiry was held by two backbenchers, the Tory John Baron and Labour's Ian Gibson, who each had test veteran constituents and wanted to get to the bottom of things. It heard evidence from veterans, widows, and radiation specialists, and produced a report calling on the government to fund a 'definitive study' into possible genetic damage to veterans from eating, breathing, and drinking radioactive particles.

By now the BNTVA chairman was John Lowe, a sailor aboard HMS *Narvik* for the first Grapple tests. He told Boniface: 'The MoD has spent years telling us black was white and producing all these studies supposedly showing there was nothing wrong with us. They inflated the figures and fiddled the results.'[8] He, and almost every other veteran Boniface spoke to, said their medical records were altered or missing. It was a version of events she heard repeatedly, with nothing to support it, and it seemed likely to be the result of paperwork simply going missing over time.

Within a couple of months, the newspaper was declaring 'victory in our campaign' as new PM Gordon Brown promised to meet the two MPs to discuss their inquiry. He said: 'The Government recognise and are grateful to all servicemen who participated in the nuclear testing programme. Their service ensured that Britain was protected during the Cold War. If there is any new information, I should be happy to look into it.'[9] A fortnight later Tory Shadow Defence Secretary Liam Fox produced Opposition pressure by saying it was time to 'be seen to be doing the right thing'. He added: 'Knowing what we know now, we owe it to them to make sure they are not suffering long-term ill effects because of what they did to support their country. If it was someone in your family you'd want to know the full truth.' In the same month, the Canadian government announced it would pay its own nuclear veterans - some of whom had been roped into the British testing - a minimum of £10,000 each. Then there was compensation for the veterans of Porton Down, who had been used for Cold War chemical and biological weapons experiments, and the Isle of Man said that it would pay a lump sum to its own nuclear test survivors. Manx vet and former Royal Engineer Ernie Moore told the paper: 'No amount of financial compensation can make up for what they did to us. I just want a simple apology.'

Boniface found a Russian nuclear veteran, Vladimir Bentsianov, who revealed he and all his comrades were given the Order of Courage, a war pension and top-up, as well as discounts on rent and public transport. He said: 'We are special people and it doesn't matter where we're from - Russia, America or Great Britain. We have risked our lives and sacrificed our health for our motherlands. Here in Russia we get the rewards we deserve. It is time you veterans in Britain were treated with respect.' He added that Russian veterans had also fallen ill with unexplained bleeding, cancers and eye problems, and their medical records were missing too.[10] The pressure worked, and in early 2008 Brown ordered his Defence Secretary to meet campaigners and pay for a genetic study. He said: 'If it is found that serving at the nuclear tests caused harm then, of course, Britain owes them a debt of honour.' Within a fortnight, the paper revealed the 'curse of Christmas Island' had struck Brown's own family, in the shape of Sqn Ldr Pooley and his questionable inquest, and it seemed like political momentum was once more in the test veterans' favour. But then it stalled.

Three years earlier, legal aid funding had enabled 1,011 nuclear veterans and widows to launch their High Court claim. Six months in, the government withdrew it. The moneymen decreed that 'the legal merit... was insufficient to justify the case being pursued at the public expense'. When the Rowland study was published, lawyers had a stronger case and agreed a no-win, no-fee deal. The claim had been ticking over, but by 2008 the two sides were destined to meet in the High Court, and somewhere in Whitehall advice was given that it was best to let the case play out. That decision meant that by the time the case was over, hundreds more veterans had died.

One of them was HMS *Diana* sailor Doug Atkinson, who joined the claim after suffering ulcers 'the size of golf balls' on his back. Suffering from cancer and given just four months to live, he told the *Mirror*: 'My chances of seeing it come to court are nil. But my wife will carry it on after I've gone.' He added: 'We were used as guinea pigs. The compensation, if we win it, doesn't mean a damn to us. What we all want, what I want more than anything, is our health. And you can't put a price on that.'[11] The following month, Boniface reported on Prince Phillip's visit to the island, the babies of the *Dunera* troopship, and documents proving Eric Denson's radiation dose. But within weeks the MoD had outlined its arguments in a letter to a veteran, which seemed jaw-dropping. 'Watching nuke tests was safer than if you had stayed at home,' ran the headline, quoting a 25-page dossier the MoD had sent to Ken Moses. The logic was based on the fact that, as there was zero radiation from the weapons, the only possible sources of exposure were natural. In the UK, granite bedrock contains pockets of radon gas that can seep out and give rise to some cancers, while there was no granite, and therefore no radon, on an island made of coral. 'If personnel who served at Christmas Island at that time had been stationed in the UK... their dose of naturally-occurring ionising radiation would have been three times greater,' claimed the report. Moses said: 'The entire report is tosh. What hurts most is that it's our own Government doing this to us.' But Labour MP Gibson pointed out the policy pivot: 'It is absolute nonsense and plainly the most astonishing bit of so-called science I've ever heard of... not long ago, the MoD admitted there was a risk and that the veterans' conditions needed serious scientific investigation. This looks like a complete U-turn.'[12]

Two months later, MoD lawyers told the court the veterans were too old to remember what had happened to them, and their memories were 'fatally and irrevocably eroded'. Dennis Shaw, who'd spent a year on Christmas Island, responded: 'If it wasn't so disgusting I would laugh - they're claiming we're senile. My memories are crystal-clear. I remember I was a fit and healthy 21-year-old, and now I have 30 per cent lung function, my daughter has a crumbling spine, and my grandson has two sets of teeth.'[13] In August 2008, the MoD had grudgingly admitted in legal papers that 159 servicemen - mostly aircrew - were irradiated. 'They WERE nuked,' the headline screamed.[14] The politicians now forced the department into another about-face, as new Veterans Minister Kevan Jones told Parliament the long-awaited health study into genetics and birth defects would take place, after all. 'I am determined to move this agenda forwards,' he said, but somehow, it didn't. Instead hearings began at the High Court, and the wheels of justice were allowed to grind the claimants down until they were, literally, dust.

Late in 2008 the phone rang again. Barry Smith had been sent to Christmas Island as a 19-year-old on National Service with the RAF, and although he hadn't

Barry Smith pictured on Christmas Island where he was barber for RAF personnel during post-bomb clean-ups. (Courtesy of Anna Smith)

seen any bombs, for a year he had cut the hair of men taking part in the post-Grapple clean-ups, and feared he had ingested radiation as a result. After he came home he had painful, undiagnosable skin problems, and his doctor told him: 'Basically, you've been cooked.' After developing a rare type of pancreatic cancer common among Hiroshima survivors, he found expert testimony from a Japanese doctor to back him up. The MoD had refused him a war pension and he was appealing. Boniface asked him to let her know how the hearing went. Just before Christmas he was told his condition was terminal, and he had to leave his deathbed to attend the hearing at the end of January. In February 2009 he rang Susie back, but was too weak to make himself heard. His wife Anna took over the call, relaying questions and answers between her husband and the reporter.

She said: 'The MoD lawyers said they hadn't looked at the evidence Barry had sent in. Barry stood up and said "Are you aware I'm dying? This evidence is not new, they've had it for four months." Then they said they didn't have time to hear the case, and it had to be adjourned... I cannot stress enough how much it took for

Barry to get to that hearing. Then to be told it was all off was devastating - and he's gone downhill ever since.' Moves were afoot to reschedule the hearing at Barry's home, but it never happened. Anna said: 'It's not about the money. Thousands of people like my husband did what they did for their country with the best of intentions. Years later, when they're ill, their country slaps them in the face. It's deliberate, and it's disgusting.'

Before Anna put down the phone, Barry indicated he had something to add. He whispered it into his wife's ear, and she repeated it. 'Barry says "Don't let them get away with it",' she told Boniface. He died a few days later. When the rescheduled hearing went ahead, Anna was refused a widow's pension on the grounds that there was no radiation at Christmas Island.

Boniface had come late to the story, seen the media coverage create political pressure and promises of change, written stories that were trumpeted as victories, only to find veterans were dying with no hope of resolution. It seemed inhumane, and with Barry's last words ringing in her ears, putting her pen down was unthinkable. The clippings show she tracked down Penney's cause of death, and his family; reported on French and Australian nuclear veterans winning official recognition; and the rulings and appeals of the High Court case. In 2010, the Redfern Inquiry ordered to investigate the theft of body parts from workers at the Sellafield nuclear plant published its report, and said at least 15 former servicemen's bodies were also harvested for heart, liver, spleen and thigh bones, some without permission. One veteran's brain was removed without his family's knowledge. Then in 2011, the long-promised genetic study was commissioned, and it was not as billed.

The Brown government had set aside £412,000 to look at the genetic impact of the tests, particularly on children, and it was delayed by the court hearings. But Labour lost power and the new Tory-led coalition government announced it would settle the argument once and for all - with a £75,000 survey. It was billed as a 'health needs audit', but managed to be both more expensive, and less rigorous, than the BNTVA's own research. It interviewed just 84 veterans in person, with another 633 completing a questionnaire. It found 83 per cent had multiple chronic health conditions, 75 percent thought it due to some extent to radiation, and a third of veterans blamed themselves for a grandchild's health problem.

Widows and descendants were intentionally excluded, no health records were checked or doctors consulted, and it was described as 'a record of veterans' individual views and beliefs'. Nevertheless it told the same story - gut, digestive and dental problems as young men, blood and immune disorders, and rare cancers as they aged. There was no real conclusion to the report, just as there wasn't an end to the veterans' fears. It said: 'For some of the veterans the mere continuation of this debate, the lack of resolution and the lack of independent and accurate information which they feel they can trust will further exacerbate their stress and anxiety and in some cases could adversely affect their mental health. This is regrettable.'[15]

A week later, the *Sunday Mirror* revealed it could have been very different. A FOI found the government was urged by scientists in 2007 to repeat the Rowland study on UK veterans, while they were still young enough for it to be feasible. The Health

Protection Agency said the research was viable, and that, as DNA can break with age, if it was to be done in the UK it must be soon. The advice was ignored. The newspaper asked Tory Veterans Minister Andrew Robathan for an interview, but he replied: 'I'm not interested. I don't want anything to do with it.' A fortnight later he had to apologise to Parliament for misleading it over the High Court case.

As another wave of redundancies hit the *Mirror* group of newspapers, Boniface - who by that point had been in Fleet Street for a decade, and journalism for 15 years - volunteered to leave, but continued reporting on the veterans as a freelance. She told how Pita Rokaratu had died at home in Fiji, a few days before the Supreme Court finally threw out the veterans' claim. A few months later, the newspaper marked the 60th anniversary of Operation Hurricane, telling the stories of Royal Engineer Thomas Wilson, who had four out of five children with serious blood and bone disorders, and Able Seaman Bill Clarkson, who developed myelomonocytic leukaemia after the Monte Bello clean-ups.

To the veterans, being ignored was normal. To Boniface it felt like a reverse from the hope she had witnessed during the court case and the end of the Labour government. In 2013, veterans marched on Downing Street with a petition demanding recognition in the form of a charitable fund to address their health problems, and former Labour leader Neil Kinnock wrote in the paper that 'our country, and our Government, has a moral duty to recognise the hazards endured by these men'. But still, junior defence minister Anna Soubry told Parliament: 'There is no evidence to support their claims, and I do not think that it would be right to set up a £25 million benevolent fund when no proper basis for it has been provided... so far I have heard no good argument to support that case.' She claimed that, as UK veterans had access to the NHS, their benefits matched the free healthcare offered to US and Australian veterans.

The idea of a fund had been mooted by Baron and Gibson years earlier, and had got stuck in the long grass. Just a few months after Soubry ruled it out, Baron stood up at the weekly Prime Minister's Questions and told his party leader: 'A third of nuclear test veterans' descendants have been born with a serious medical condition. Our cross-party campaign seeks recognition, not compensation, and includes a Government ex-gratia payment into a charitable fund to help those in need. Will the Prime Minister... clear the logjam; recognise our veterans; and resolve this shameful chapter in our nuclear history?'

David Cameron's reply was brief, but had a different tone to his minister's. He said: 'I am happy to tell the house that the government recognise and are extremely grateful to all the service personnel who participated in the nuclear testing programme. We should be in no doubt that their selfless contribution helped to equip the UK with the deterrent that it needs.' Baron greeted it as the first formal recognition of the veterans, and it was followed by the announcement that a £25m fund would be established.

But there were two catches. The Aged Veterans' Fund would be open to bids from multiple organisations, so the money would be shared. And none of it could be used to meet or investigate the health problems of test veterans' estimated 155,000 descendants, about whom survivors were most worried, and who were now the focus of the campaign. Cameron never addressed Baron's point that the children and grandchildren were suffering too.[16] The *Sunday Mirror* reported on 'Nuke vets: the new generation' and retold the story of Shirley and Eric Denson, whose baby great-granddaughter was now suffering from a duplex kidney. That was followed by the case of ex-RAF serviceman Bob Fleming, and what had befallen his family after his return from Christmas Island.

'Even as odd illnesses began to strike his children, Bob and his wife Jean put it down to bad luck. But the believing has long been over. Because that "bad luck" has left a horrifying 14 of the couple's 21 descendants - 66 PER CENT - with birth defects and deformities,' reported the paper. 'The figure is 33 TIMES worse than the national average.' Jean said: 'Every time there's a pregnancy you say, "Oh blimey, is everything all right?" You have to find a way to live with the worry because you've no choice.'

Bob lost all his teeth in his 30s, but their first child was born healthy. Next came a daughter who developed thyroid problems, breathing difficulties, and lost her teeth. Of her four children, two had heart problems, a third had severe asthma, while her grandchildren's health problems include missing or blocked tear ducts, arrhythmia, a deformed pelvis and no tooth enamel.

Bob Fleming pictured outside his house in Denver, Norfolk, with granddaughter Sophie Ward, then aged eight, to mark Armed Forces Day, in June 2014. (Courtesy of the Fleming family)

Bob's last child was another daughter. The paper reported she 'has only one working lung, and extra bones in her knuckles and elbows'. Her jawbone was crumbling from widespread osteoporosis, her teeth were coming out, and she also had blocked tear ducts and saliva glands. Doctors had asked if she had ever been exposed to radiation, as it could explain her problems. Her children's health problems included another pelvic deformity, miscarriage, and gastric problems and breathing difficulties. 'How can anyone argue with all this?' asked Jean. Bob said: 'I wish I had never gone to Christmas Island... they wouldn't do it now, so why can't they admit they shouldn't have done it then?'

The same story was being uncovered in Australia, where Frank Walker published his book *Maralinga* and found veterans who had severe radiation sickness after the tests, brain tumours, and children who never grew teeth or hair. The wife of one he spoke to had six miscarriages, three living children with congenital defects, and a son who developed acute myeloid leukaemia at 41.

These were not unproven fears. Researchers in Europe and the US had shown that cells which seemed to have survived radiation unharmed could create 'daughter cells' showing delayed damage, a phenomenon called 'genomic instability'. But while it had been proved with cells, it was harder to do so for a complex organism, which sheds and creates cells all the time. Little work has been done on the topic, outside of cancer patients or those exposed in Japan, as medical ethics forbid exposing someone to radiation without a therapeutic benefit, and nuclear accidents that could create study subjects are something people try to avoid. Most scientific research is funded by governments or major corporations, and while grant mechanisms strive to make it as fair as possible, they can choose to simply not fund what looks inconvenient. The wealthiest nations in the world are, in general, those with the most scientific research, and the most to gain from either nuclear weapons or nuclear power, so there is an incentive to not look too closely at the possibility of genomic instability in test veterans' families.

Those nations which have exploded nuclear weapons, with the potential legal liabilities that could spawn along fallout plumes, have had a moral or political imperative to investigate the results. Yet in 1980, a Congressional committee reported the US monitoring programme on the health effects of its own tests in Nevada and the Pacific 'was inadequate and, more disturbingly, all evidence suggesting that radiation was having harmful effects, be it on the sheep or the people, was not only disregarded but actually suppressed'.[17]

In post-war Japan, the US agreed to fund the Radiation Effects Research Foundation, which is still doing work today on the long-term impacts of Hiroshima and Nagasaki, including on the lifespan and cancers of survivors' children. But its birth defects study ended in 1954. It concluded that 'research thus far has not indicated any genetic effects in A-bomb survivors' children'. It is still quoted 70 years later by the MoD and ministers as proof the veterans have nothing to worry about, although few in Japan experienced the same size or type of weapons, nor the same forms of exposure, as servicemen in the testing grounds, and no modern techniques or knowledge have been reapplied to the question.

Michael Watson was on National Service when he was sent first to Maralinga, and then Christmas Island, with the Royal Engineers. He told wife Margaret he wore just shorts for the tests, and that he swam in the sea and ate the fish. After his return, the couple suffered nine miscarriages. They had three surviving children: one boy is autistic, a second son is infertile, and daughter Sharon was born with S-shaped bones in her legs.

She told the *Mirror* about a lifetime of pain, and how some of her teeth grew sideways into her gum. She suffered 12 miscarriages, and her two surviving sons have severe co-ordination problems and tooth abnormalities. As she approached middle age, Sharon's leg muscles could no longer cope with her deformity, and regularly went into spasm, causing falls. She endured years of surgeries to break, pin, and reshape every bone in her legs. Her husband gave up work to be her full-time carer, and the family were surviving on benefits.

The two women put it all down to bad luck, until Michael was diagnosed with cancer in 2007. Sharon said: 'He never spoke about it for years because he'd signed the Official Secrets Act. Then when he was diagnosed he started to talk. He showed us pictures and described the flash of the bomb. We saw stories in the paper about people being ill after the tests and it started to make sense.'

She added: 'It wasn't the men's fault. They didn't know what was happening. I'm proud of what my dad did but I'm not proud of my government. Those of us with problems need a little extra help. We can't live on benefits and I need to adapt my house for my disability. My husband wants to work and I want to stand on my own two feet. Why are we dismissed like we're malingerers, when all we want is to be acknowledged?'

Another descendant was Steve Clifford, who developed a rare adult leukaemia when he was just four years old. Doctors told his parents it could be due to exposure to radiation. He was given a blood transfusion and six months to live. Steve said: 'The family were certain it was because my dad Graham was a cook at Christmas Island in 1957. He used to cook and eat fish the men had caught in the sea, between the bombs. He said once he cooked a lobster for the officer's mess and when the officers jokingly tested it with a Geiger counter it went off the scale. But they ate it anyway, because they didn't know the effects.' When he came home Graham had an undiagnosed growth removed from his abdomen, but otherwise enjoyed good health. Steve survived his cancer but grew up with constant pain in his joints, undiagnosed lumps on his calves, and serious back pain by the time he was 20. In his 30s bones in his shoulders began growing spurs which affected his movement. He had to leave his job as a brewery engineer in 2010 and developed fibromyalgia and sciatica, as the bone problems put pressure on his nerves. One day, several teeth fell out while he was in the supermarket, for no obvious reason.

He and ex-wife Dawn had two sons and two miscarriages. The eldest boy had hypermobility in his joints, which means in simple accidents he suffers severe injuries. In 2014 a rugby tackle turned his leg back to front and he needed surgery to reattach the tendons. Steve said: 'When you went to the doctor in the 1970s and mentioned Christmas Island they said there was no link, but today they say "Oh that might be it". That's how formidable the MoD's policy of denial has been, but now they're swimming against the tide. I worry my boys will develop the same problems as me when they get older. I want them to know the facts so they can make the right choices.'

In 2012, a war pension test case brought together 14 radiation-linked claims for five survivors, three widows and six deceased veterans of Christmas Island and Maralinga. Between them, they reported cancers of the blood, pancreas, and bladder, heart disease, diabetes, kidney disease, skin ulcers, and cataracts. One of the claimants was Barry Smith's widow, Anna. The case ground on through appeals until 2016, when the High Court ruled that, while many of the conditions were radiogenic, there was no evidence that any man had a big enough dose to link service to injury, and threw them out.

In those claims - and all others - veterans had to ask the MoD to provide evidence of radiation exposure with which to incriminate itself. They never believed that what was produced for legal hearings was the full truth. And shortly after the judgement, there was proof the MoD was not acting in good faith.

Documents were uncovered which revealed the government had been prepared to pay out war pensions to all test veterans 28 years earlier, but it was blocked by the MoD, and forgotten about by ministers. In February 1988, after the first NRPB report, health officials were working out how to apply its findings on leukaemia to the war pension claims which were coming in thick and fast. Dr Owen Eggington, chief medical advisor at the Department of Health and Social Security, reviewed the study and wrote a memo to his bosses. He said they could refuse all claims for solid cancers, but 'the situation is different' for some types of blood and bone cancer. He wrote: 'The onus of proof is such that the Department has to show beyond reasonable doubt that the conditions claimed are NOT due to service. I do not believe that we can do this and thus all these cases should be allowed.'[18]

He added: 'There is, in my opinion, no doubt that the [NRPB] report constitutes reliable evidence... clearly our decision regarding these few cases may cause problems with the Ministry of Defence and we will have to explain our stance... there could possibly be a problem regarding the public in general and the media in particular who may see our decision as two government departments working in opposite directions.' The memo caused fury in the MoD. Mike McTaggart, then Second Permanent Secretary, rang his opposite number at the DHSS in what sounds like a hissy fit, accusing them of going behind the MoD's back, briefing ministers, and not consulting them. The papers show Eggington and his team were bemused, talking about 'a serious difference of opinion between our doctors and the MoD' and saying McTaggart was behaving 'quite unjustifiably'. They even raised the question of whether government officials were being pressured to break the law by the MoD.

An internal note says: 'I am not sure how we resolve things... the problem is basically a medico-scientific one of how the report should be interpreted. As Dr Eggington says, if our doctors feel it raises a reasonable doubt, the law requires them to make an award. However, it would clearly be extremely embarrassing if this Department were to follow this line while the MoD are maintaining that the report does not raise a reasonable doubt.' The whole affair was due to be discussed in Parliament, with a one-line motion tabled at the end of that month calling for a full debate on the issue of war pensions for nuclear test veterans, and talk of a possible judicial review to decide whether the government's position was legal. The papers include a briefing note from McTaggart, advising the Leader of the House - the government representative organising Parliamentary business - how to handle it, and pointing out a minister had already said there was no need for judges to get involved. Somehow, other Parliamentary business knocked it out of the schedule. He added a gleeful hand-written annotation - 'It was not reached because of lack of time. Unless it is raised again we can forget it!'

Two decades later, the creation of a benevolent fund meant changes for the BNTVA. From a loose association in which every veteran had a vote, it became a charity with trustees, and only they had a say over its direction. It bid for and received £1m from the Aged Veterans Fund, and established the Nuclear Community Charity Fund (NCCF) to administer grants for home adaptations and care equipment, as well as to support research and memorialisation. One of their first projects was to fund a new centre looking at the health effects of radiation at Brunel University in west London including, finally, a genetic study.

It promised a more detailed analysis, and for the first time to consider the DNA of family trios - father, mother, and child – to see if genetic damage had been passed down, and from whom. But it was too little, and too late. The veterans were a decade older, with a higher likelihood of DNA damage due to age, as well as having their numbers thinned. The study could use only those who had not had radiotherapy, which ruled out the worst-affected. It was difficult to find 50 family trios, the minimum number required, in whom a father and mother in their final years were cancer and dementia-free, and capable of consenting, along with a child.

In early 2018, Boniface reported the university had found only two. Maralinga veteran Jeff Liddiatt, once chairman of the BNTVA and now taking on that role at the NCCF, told her: 'We're getting thin on the ground. The MoD has spent decades kicking this issue into the long grass and refusing to do this sort of research in the hope that we'd all be dead before long. Now we have paid for the research ourselves we just hope it's not too late to get the answers we know are there.'[19]

After a Brunel press conference, Boniface was approached by the new BNTVA leader Alan Owen. He had become chairman almost by accident. One day, arriving at a garden centre, he saw a man outside with a collecting tin for the BNTVA. That man was Doug Hern, a long-time BNTVA figure who became its historian. Owen told him that his father had been at Dominic, and Doug replied: 'Either he is ill, or he's dead.' Owen had been unaware of the recognition campaign, but thanks to Doug's inspiration became a trustee and then chairman. A software developer with his own company serving the charity sector, he had a drive which some of his predecessors in the role had lacked, through no fault of their own. At the first reunion after taking over, he asked the veterans what they wanted him to focus on. Colin Moir, a former Royal Engineer at Grapple Y and Z, stood up, took the microphone, and said: 'I think we should ask for a medal.' The idea was popular with the veterans present, and Owen asked Boniface for the *Mirror's* support. He also

Former sapper Colin Moir, of Hatton, Aberdeenshire, was ordered to witness and repair damage from five of the Grapple blasts in 1958. In 2018 he suggested asking for a medal, which was to prove pivotal to the battle for recognition. (Susie Boniface)

dangled a tempting carrot: a group of veterans were going back to Christmas Island for the 60th anniversary of Grapple Y. Would she like to go with them?

By now the mother of a young child, she could justify neither a month-long trip to the end of the world and back, nor a jaunt into Ground Zero. Instead Boniface approached Alison Phillips, who had just taken over as editor of the *Daily Mirror*.

Doug Hern pictured with his two daughters at home in Lincolnshire in the 1970s. (Courtesy of Alan Owen)

Its coverage had become lacklustre, struggling with a lack of awareness in the constant newsroom churn of staff. Phillips was, like Stotty, a campaigner first and foremost, and when Boniface brought Owen and the indefatigable Shirley Denson into the newsroom to discuss the medal campaign, she was hooked. The *Mirror* and the BNTVA planned a publicity blitz that would get the veterans back on the news agenda, for press, public, and politicians alike.

At the end of April the *Mirror* ran a two-page feature on the four veterans who were returning to the Pacific. They included Robert McCann, who had worked on the clean-ups and Grapple, as well as Ray Carbery, a former mechanic whose job it was to fix the fridges that kept the beer cool on Christmas Island, and later developed constant bowel problems and had a severely autistic son. 'We watched the planes sent into the mushroom cloud come back down, and the scientists took off samples packed in flasks. They had white protective suits, helmets, gloves. We were 20 yards away in shorts, no-one cared about us,' he said.[20]

Also going back was Leslie Hawkins, once a 19-year-old RAF technician at the Grapple tests at both Malden and Christmas. 'Some of the islanders have similar medical problems to us, but they were told if they asked the British government for compensation all the UK money would be taken away,' he claimed. Ex-sapper Ron Watson lost all his teeth within two years of leaving Christmas Island, and a decade later had a thyroid tumour removed. He wife suffered two late-term stillbirths, and they later divorced. 'My thyroid is probably in Porton Down,' said Ron, who as BNTVA treasurer was leading the delegation. 'I don't dwell on it, but there should have been other children. We were guinea pigs, whatever the MoD says.'

A month later, Phillips ran a second feature, demanding Britain 'Honour Our Nuke Heroes' with a medal. 'As boy soldiers in 1958, thousands of this country's fittest young men were ordered to a tropical paradise to take part in the biggest military operation since D-Day. Today they have been honoured by the impoverished islanders who still live in the testing grounds - while in the UK successive governments continue to obstruct calls for official recognition,' the paper thundered.[21]

After going freelance, Boniface had reinvented herself as a columnist and broadcaster, with a high profile on social media. It brought her the ears of people who would never usually pay much attention to mere news reporters, and she decided to exploit it. She persuaded Labour's deputy leader Tom Watson to back the campaign, and he vowed: 'Their service was unusual and of crucial importance to Britain's place in the world. I will do whatever I can in Parliament to ensure they get the medal they so justly deserve.'

The next day he was filmed in his Parliamentary office interviewing ex-RAF clerk John Ward. John said: 'You felt the bomb before you heard it – the heatwave came across you, then there was a loud crack and rumble like a million degrees of thunder. We saw this mushroom cloud rise from the sea, and if you said you weren't scared you were a liar.' But he blamed himself, and his service at six Grapple bombs, for his daughter's problems. Denise had most of her teeth removed as a child, suffered jaw damage, and had 14 years of fertility treatment to no avail. Aged 56, her pancreas stopped working properly, and she developed calcium growths on her shoulder. Her brother lost his twin in a partial miscarriage before he was born, and later had a tumour removed from his kidney.

Denise said: 'I never know what I'm going to wake up with next... to be perfectly honest it feels like I'm drowning.' Her father added: 'I was proud to serve my country but they used us as guinea pigs, or at the very least just didn't care.' Afterwards, the MP wrote for the paper, saying Labour's failure to help the veterans when in power was a 'stain on our record that for so long [governments] asked these men for proof of what was done to them, when all most of them wanted was our thanks... a medal would go a long way to healing some of their wounds'.

mirror.co.uk

DAILY Mirror

FIGHTING FOR YOU

Thursday, May 31, 2018 75P

Hot to trot: Strictly's new stars

SIZZLING DANCERS Graziano Di Prima and Luba Mushtuk

SEE PAGE 9

THE PROOF

Documents reveal British nuclear test crews WERE used as human guinea pigs by their own government...

DAMNING Papers revealing experiments in deadly nuke cloud

In view of the fact that the pilot appears to be very much more exposed than any other crew member, the 1.8 maximum difference (15.8 r to 17.0 r) suggests that the perspex canopy is a far more efficient shield than had been suspected. Nevertheless, the considerable higher general level of radiation to the head would appear a possible hazard in relation to the sensitive eye tissue. But as it appears beta radiation is largely absent from the cabin, this could perhaps be over emphasised, as experiment has suggested human eye tissue to be less sensitive than that of the animals which have been used experimentally.

TRAGEDY RAF ace Eric Denson

EXCLUSIVE

BY SUSIE BONIFACE

BRITISH pilots were deliberately exposed to radiation in Cold War nuclear experiments, documents have confirmed for the first time.

Widow Shirley Denson obtained papers that show her husband was ordered to fly into the cloud of an A-bomb in 1958 in the Pacific. He later killed himself.

It shatters decades of MoD denials that men were used as guinea pigs. Shirley said: "It's wicked."

FULL STORY: PAGES 4&5

The *Mirror* splash with the first evidence of radiation experiments conducted with British troops, and the 'initial experiment' on Eric Denson's plane during Operation Grapple Y. (Reproduced with the permission of Reach plc)

Two days later the *Mirror* splashed on the revelations about Eric Denson's experiment. The front page screamed: 'THE PROOF - documents reveal British nuclear tests crew WERE used as human guinea pigs by their own government'. Shirley was quoted, saying: 'It's absolutely wicked. It's evil. To see it in black and white after all these years took my breath away. It seems our government used and

6. Sampling Effort on D-Day

only normal sampling Canberras could be used for sampling on Grapple-Y.

By 0600 hours local on D-Day, five aircraft had been fitted with Mark 3, 9 and 10 filter elements, drums, evacuated and spiked gas bottles, cabin air sampling filters and integrating dosimeters. All external equipment had been dried and was covered with polythene bags containing silica gel packets to combat moisture. These bags were removed just before the aircraft taxied from dispersal. The detailed operational procedure is given in a No. 76 Squadron report).

The cloud did not rise to the expected altitude and the sampling procedure had to be re-assessed by the sampling controllers ( ) in Sniff Boss (aircraft WJ.754).

Sniff Boss first reconnoitred the cloud at 1025 hours local and made two passes through cloud stem at 42000 ft, each of about a minute duration. At 1049 Sniff 1 (WJ.757) entered the cloud on a first run at 46000 ft approximately, on a heading of 350° and took no indicated integrated dose. Sniff 2 (WH.980), was sent in at 1055½ hours, made one pass lasting 3½ minutes at 52800 ft indicated and received an indicated maximum dose of 10.6r on returning to base. Sniff 2 was then sent in again at 50000 ft at 1109 hours for five minutes, and subsequently once more for a further six minutes, receiving in all an indicated dose of 8.3r.

Sniff Boss contacted "Cleaner" (Decontamination Control), to enquire whether the sample from Sniff One and Two was sufficient but as Sniff Two had not returned, Sniff Boss was advised to send in Sniff Four, Sniff Three having become unserviceable with a non-functioning Mark 3 duct actuator. Sniff Four made three runs between 1150 and 1223 hours and returned to base at 1250 hours.

~~SECRET~~

-6-

A page from the MoD reports of air sampling at Grapple Y, produced after a Freedom of Information request. Sniff Boss, plane WJ754, was flown by Terry Gledhill with Joe Pasquini as navigator. Eric Denson was in Sniff 2, WH980, and was sent into the cloud three times. Document reference FOI2018/06695.

abused its own men. It makes me furious to think it was done on purpose, that my Eric mattered so little to them.' Owen added: 'This is the first time in all our years of campaigning we have ever found evidence this strong. Our members always believed they were guinea pigs and this appears to prove some of them were, at best, collateral damage in horrifying experiments.' Watson toured TV studios calling it 'a shocking document the MoD cannot wriggle out of... we need answers about what experiments were conducted... the Defence Secretary should come to the Commons and issue an unqualified apology to Mrs Denson.' The Labour frontbench rowed in behind him, with the shadow defence and health secretaries echoing his call.

But wriggle the MoD did. It released a statement, saying: 'It is not true to say these men were subject to an experiment to look at the effects of radiation. The British nuclear testing programme contributed towards keeping our country secure during the Cold War and regular health checks were conducted throughout. The NRPB has carried out three studies of nuclear test veterans and found no valid evidence to link this programme to ill health.'

The campaign developed a head of steam, driven in no small part by Watson's involvement. An MP who'd led high-profile battles against both Press intrusion, and child abuse allegations which later turned out to be fake, he was widely recognised by the public, while much of the media loathed him. As deputy leader he was understudy to Jeremy Corbyn, a peripheral figure who'd been propelled to the leadership by a Hard Left takeover of the party. They didn't agree on much, but now had test veterans in common. Corbyn was an old school, anti-nuke Leftie, while Watson was a former Armed Forces minister in need of a campaign that wouldn't bring Press hatred.

On June 6, 2018, he organised an event in a side room off Westminster Hall, the 1,000-year-old public space in Parliament. It is where Sir Thomas More stood trial, Nelson Mandela addressed the Queen, and Charles II held his Restoration banquet. The event was advertised with an email to MPs asking them to pop along and show support. As the *Mirror* later reported, 'You're lucky if more than six MPs find the time to locate the right corridor and poke their head around the door to say hello'. After eight had been and gone, the photographer remarked: 'Well, that's your lot. Pretty good show.' Boniface described committee room W2 as 'a stained, unloved and rarely-vacuumed room about 20ft square', accessible only by navigating a retractable safety barrier, and a corridor filled with filing cabinets. And into this, after Prime Minister's Questions, filed three peers, 40 MPs, half the Shadow Cabinet and a government minister. Although the room was filled from door to window, standing room only, they all met Owen, Shirley, Doug and his wife Sandie, as well as comedian and historian Al Murray, a famous face and a friend of Boniface, roped in to draw the crowds.

Doug Hern was a 21-year-old Royal Navy chef when he was ordered to stand on a spit of sand and watch the bomb. He said: 'Afterwards I was detailed to pick up the carcasses of birds and fish killed by the blast, and we just wore shirts and boots. We'd go out on a boat, and if there was 4ft of water, 2ft of it was dead birds and fish. We just scooped them out and took them back. Other times the scientists sent us out in shifts, fishing for live ones so they could dissect them. We never had any protective gear. I was on the island for a year, and we never saw a safety officer. We used to go to Ground Zero to catch crayfish off the reef, drink beer and barbecue them and no-one ever stopped us.'

Afterwards Doug lost all his teeth, had skin problems and diabetes, and developed bony spurs growing from his ribs into his chest. He also fathered two daughters. Gill died at the age of 13 from Cushing's Syndrome, an extremely rare kind of cancer of the adrenal gland more common in horses and dogs than in humans. It causes excess hair growth and by the end of her life, Gill had to be shaved twice a day. His other daughter was unable to have children.

Doug Hern was a navy chef for the Grapple tests, and later nursed his 13-year-old daughter Gilly as she died from adrenal cancer. He suffered extensive health problems but served as BNTVA historian for many years. He died in August 2023, before he could receive the medal he had done so much to campaign for. Pictured with his wife Sandie at the All Tests Reunion she organised every year. (Courtesy of Alan Owen)

Doug said: 'I'm lucky in that I'm fairly robust. But I have watched my girls suffer. It shouldn't be us that has to prove what happened, the Government polluted our bloodline. I held my girl in my arms and prayed for her to die, and that's something I'll carry with me to the end of my days.' Doug said Gill was photographed in hospital almost every day during two years of treatment. When he asked for them, he was told no records existed.

Birmingham Yardley MP Jess Phillips listened to Shirley's story and called it 'premeditated murder... a total dereliction of duty on the part of the MoD and it's not the responsibility of veterans to prove the link. The link is obvious'. For the first time, a serving government minister also supported the campaign. Tory Suella

Braverman, then a junior Brexit minister and a future Home Secretary, said: 'There would be a lot of support for this. It's really important to hear their stories of decades-long struggle and what their families have had to put up with. I think there would be a lot of support for this campaign and Parliamentary interest if we can get this heard.'

Boniface wrote: 'I thought we'd hit the peak. The room was jammed, there were MPs queuing down the corridor, I was pushed up against the back wall trying to get a quote from someone and was just thinking three reporters would not be enough to keep up with all this when I glanced at the other end of the room and saw Jeremy Corbyn having his picture taken with Shirley. One of Tom's aides was stood 6ft from me, maitre d'ing the doorway. We made eye contact over a sea of heads, I gaped and pointed, he gaped and nodded. This was the first time that a party leader had even MET a veteran. And here was Corbyn, giving them a hug, joking with Sandie about how she liked men with beards, and actively backing the entire campaign.' Corbyn was merely being consistent. In 2015, when running for leader, he'd been the first candidate to give unequivocal support to the veterans. His position meant he brought many of his MPs with him, and the three old campaigners were still being approached for photographs as they left Westminster Hall several hours later. 'We've never had a response like that in all our years of campaigning,' said Doug. 'I just can't believe it.'

It was followed by a request for Shirley, with Al Murray, to discuss it all on Sky News. A MoD press officer rang Boniface - the first time the calls had gone in that direction - and offered a comment. He said: 'We are grateful to all those who participated in the British nuclear testing programme which contributed towards keeping our country secure, and we carefully consider every request to consider their extraordinary accomplishments.' It was the first time the government had referred to the veterans, or their service, as 'extraordinary'.

One of those who came to the committee room was a Tory backbencher called Sir John Hayes. Not well-known to the public, he was the sort of politician who knows which ear to whisper into, and in what corridor that ear could be found. He was also Doug's local MP, the patron of the BNTVA, and as a junior minister in 2014, had worked behind the scenes on the benevolent fund that was set up. Within a fortnight, he had secured the campaigners a meeting in the MoD. He stood up in the Commons and asked the Defence Secretary, Gavin Williamson, if he would meet 'those brave people' and discuss a medal. Williamson, whose lack of any military background made him grit in the MoD's oyster, said he would 'be honoured'.

When the day dawned, the indomitable Shirley - who could barely walk due to ulcers on her feet - strode into the room, and in front of half a dozen aides and officials, looked the Secretary of State for Defence squarely in the eye, and said: 'So. You're the man who killed my husband.' She wasn't strictly accurate - Williamson was born two months before Eric took his own life, and 18 years after Grapple Y. But it made the point that someone in that job was ultimately responsible, and it set the tone for a meeting in which one of the most powerful men in the country sat and

Doug Hern, left, his wife Sandie, centre, and Shirley Denson, on the right, pose for a picture with Al Murray after their successful day in Parliament. Boniface stands next to Doug, holding a picture of Eric. (Author's own)

listened, without argument, to people who knew first-hand the effects of nuclear weapons. Shirley said afterwards: 'You could hear a pin drop as I told him about Eric's experiment and the effect it had on our family.' He seemed convinced, and afterwards he ordered fresh scientific research and ordered a medal review to look into the 'exceptional circumstances' of service at the tests. Williamson told them: 'It is time to take a fresh look at this.'

Owen recalled: 'Shirley took in a picture of Eric and showed everyone, security guards, literally everyone. She was amazing. I told him the veterans did their duty and said, "Now it is time for you to do yours". We hoped for progress on the medal, but the new study is a superb bonus, beyond our expectations.' Williamson told the *Mirror*: 'I was honoured to meet with Shirley Denson and Alan Owen... We must never forget the important work of our nuclear test veterans in contributing to keeping our country safe during the Cold War and the effect that still has today. I have commissioned a new study into their health and well-being and I hope the results of this will offer certainty and clarification for this group of brave individuals and their families.'

Unbeknown to the campaigners, just two days earlier the Warhead Director of the MoD's nuclear weapons department had written to the National Archives, demanding it hand back 60,000 records, many relating to the tests and with the reference code ES. Although these had been in the public domain for decades, without any concerns, he said they now presented a risk to national security.

Perhaps as a result of the renewed campaign, the AWE had been reviewing the records, and the letter said 'the closer we examine the ES series, the more concerned we become over the proliferative potential of the records'.[22] It was followed by a request a few months later for a second batch of files, the AB series. The *Guardian* later reported that 'a vast cache of material dating from 1939 to the 1980s... has been unexpectedly withdrawn.' Academics who were in the middle of research were outraged. Jon Agar, a history of science professor at University College London, said: 'We would be alarmed as historians that it has been taken out of public view. These are important records... a couple of days ago a PhD student noticed that everything in the record is coming up as temporarily retained. We are all scratching our heads. It is all a bit mysterious.'[23] No date was given for their return. The timing seemed suspicious, but there was nothing that could be done about it.

It had been just 44 days since the *Mirror* launched the medal campaign, and it seemed like celebrity and cross-party political support, coupled with media attention, was producing results in double-quick time. Veterans were elated, and the BNTVA and *Mirror* campaign re-energised. But the problem with getting into the MoD was that, now, veterans were reliant on it. To forestall any problems, Watson demanded the research look into the health of children and grandchildren, and urged Williamson to take personal charge. A month later, a Parliamentary petition hit 10,000 signatures and triggered an official response from government. It said: 'The Government continues to recognise and be grateful to all Service personnel and civilians who participated in the British nuclear testing programme. Their selfless contribution ensured that the UK was equipped with the deterrent we needed during the dangerous years of the Cold War... the MoD has asked... if the Committee on the Grants of Honours, Decorations and Medals would consider looking afresh at the previous information, and examine any new evidence presented by campaigners.'

On the face of it, the campaigners now had a Defence Secretary, the Labour frontbench, the Tory backbenches, a fresh health study, and a medal committee devoted to their cause. But all was not quite as it seemed.

Within a month of the meeting with Williamson, it became clear the MoD would not blindly follow orders. John Ward had a letter from officials saying a medal would hinge on proof the men were harmed by radiation. Watson pushed back, demanding to know if the MoD intended to create 'first and second-class medals, for first and second-class heroes'. He said: 'It would be unprecedented for a medal to be linked to injury like this. What would it mean for heroes of D-Day - should they not wear a medal if they can still walk?' A minister was pushed out to apologise for 'the awkward language', but the MoD refused to confirm who was doing the medal review. The military medal committee had been formed and then disbanded in 2013 to consider a number of claims, including that of the nuclear veterans, and thrown out all but one of them - a gong for members of the Arctic Convoys, who protected sea transports during the Second World War. At the time, Russian leader Vladimir Putin had just awarded one to his own Arctic veterans, and doing the same in Britain paved the way for his state visit to the UK the following year. It proved the medal would come only if it was politically necessary, and it wasn't. Five months after seeing Williamson, there was still no study agreed, and no committee formed.

It was not until April 2019[24] that the Queen - as the official source of all honours - was finally asked for, and approved, a medal review. Dr Charles Winstanley, a former army major who saw active duty in the Middle East and Northern Ireland, was appointed chairman of the new Advisory Military Sub-Committee. The AMSC would consider applications, and pass its decision to the main Honours and Decorations Committee, which would make a recommendation to the Queen. But within days Williamson was sacked after being accused of leaking state secrets, and the veterans saw it as the MoD moving against the man who might have helped them. The medal committee formally opened for business a few months later, and the BNTVA lodged an application in August, supported and reviewed by Hayes and Watson, who both had experience of the submissions process for honours. There would be no decision for an inexplicable 16 months.

The Labour Party had woken up to the political capital that could be gained by supporting the nuclear veterans. In September 2019, ahead of the annual party conference, the *Mirror* splashed on 'Justice For Nuclear Veterans' as the party pledged to give £50,000 a head to every survivor. Boniface had taken a call from a Labour advisor at home one Sunday afternoon, and could barely believe it. The £75million cost had been signed off by Corbyn and his entire Shadow Cabinet. The party's foreign spokesman Emily Thornberry said: 'While I had read their stories in the *Mirror* for years, it was only when I talked to them [in Parliament] that I truly realised the pain they had been through, especially when their children and grandchildren began to suffer health problems... It is not just our role as a party to fight against the injustices we see today, but also to correct the injustices of the past. And justice for these veterans cannot wait a second longer. We don't need science to tell us what is right under our nose.'

A day later she gave a speech praising the *Mirror's* campaign, and promised to bring an end to it. She said the money would help with veterans' medical needs and 'give them the support and comfort they deserve in their old age'. She added: 'We need a PM at the despatch box not just making good on this commitment we've made, but issuing a formal apology to the nuclear test veterans and families on behalf of the whole country.' An embarrassed Boniface was forced to accept a round of applause from a conference hall of party activists who normally loathed tabloid journalists.

Doug was among 105 veterans who watched the announcement live from their annual reunion. He said: 'It's a good thing Labour has done this and we hope everyone else will follow suit. When I look back now, my daughter was treated more like a medical specimen than a patient. Nothing can bring her back, but this is long overdue.' Two months later the paper reported one veteran was dying every week while waiting for news.

Derek Redman was on the regimental boxing team and rated A1 fit, but died of 'diabetic coma' within a few hours of entering Ground Zero at Christmas Island in 1958. Such a cause of death cannot be confirmed without a post mortem, yet his inquest was told there wasn't one. (Courtesy of Donald Redman)

Many veterans had died decades earlier, but that was no reason to forget them. In March 2020, the *Mirror* revealed what it claimed was a continuing cover-up over the death of Lance Corporal Derek Redman of 38 Corps Royal Engineer Regiment. Derek, a former National Serviceman, loved the army so much that he signed back up when he was demobbed. He won regimental medals for boxing, athletics and tug-of-war. He scuba dived and was one of the healthiest lads his unit had when, in 1958 aged 27, he was sent to Christmas Island for Grapple Y. His medical check-up before leaving rated him 'A1' fit. He was expecting to return home to his sweetheart, Beryl, with a fine tan and some tall tales, yet a few months later he was gone. In the course of a single day he developed 'diabetes of sudden onset and great severity', slipped into a coma, and died.

His brother Donald said: 'My brother wasn't diabetic. He was built like a brick outhouse and never sick a day in his life. My mother never stopped grieving for him, he was a lovely lad. We've always believed the radiation killed him. The MoD used him in their experiments and then swept it under the carpet.' Derek's former comrades confirmed that, in the week before his death, he had been ordered into the blast zone without protective clothing. Terry O'Keefe said: 'He was a driver and the day before he died, he and I took a trip to one of the lagoons on the island to go swimming. Our truck got stuck in the sand and we had a right job getting it out. I remember him pushing it with his bare hands. When we got back to camp he said he didn't feel well and wouldn't come for tea. The next day he was dead, and I was ordered to form part of the firing party for his burial at sea. I couldn't make head nor tail of it, because he'd been manhandling a truck only the day before.'

Fellow sapper Dave Whyte was also in the firing party. He knew Derek as 'Dal', and said after the test he repeatedly drove scientists 'to the forward area, where radiation levels were extremely high'. He said: 'He was not given any protective clothing or a respirator to prevent him receiving a large dose of radiation. He had done this journey on numerous occasions and one evening he said he wasn't feeling very well when he went to bed. He felt worse the following morning, and asked a friend to report his condition to the squadron sergeant major.' Derek was found unconscious in his tent on the morning of May 6, 1958, and was transferred to the Christmas Island infirmary where he died in the early hours. Derek's coffin was taken aboard HMS *Narvik* for the burial-at-sea ceremony. But the coffin floated. Sailors were ordered out in a dinghy to drill holes in it.

Newspaper archives show Derek's parents did not accept the official account. John Redman, a postman from Brighton, told the *Daily Herald*: 'I don't believe that my boy, a fit, tough lad, could develop diabetes and die within 24 hours.' Mum Ellen added: 'Why did they bury him so quickly at sea? It looks as though there was something to hide. I am sure we haven't been told everything. I can't help thinking there was some sort of accident of the H-bomb tests that affected him.' Derek's commanding officer told the paper that 'he had been working with the scientists for about a month... when he returned to the regiment, everyone noticed he looked rather thin.' The family's MP asked questions in Parliament. Julian Amery, Under Secretary of State for War, told the Commons: 'He was suffering from diabetes of sudden onset and great severity, and he died in the RAF sick quarters early next morning. There had been no sign of serious illness when he was treated for a minor ailment on 4th and 5th May. I am advised that severe diabetic cases of this kind are not unknown, although they are comparatively rare. L/Cpl Redman was buried at sea... because there is no consecrated ground on Christmas Island. He had been employed in charge of a power station. His work was not directly concerned with the nuclear test explosions... I can reassure the House that there is no medical connection whatsoever between this soldier's death and the nuclear test explosions.'[25]

Boniface explained: 'Professor Partha Kar, a consultant endocrinologist and adviser on diabetes to NHS England, says it's vanishingly rare someone who appears fit and healthy, with none of the usual symptoms of diabetes - weight loss, frequent urination, dizzy spells - would develop it overnight. And he told me the only way to be sure it was diabetes is with a post mortem to examine the pancreas. Its cells make insulin to regulate blood sugar, and when someone has diabetes it is these cells that fail. If Derek had been born with Type 1, his immune system would have attacked those cells, probably since his youth. If he developed Type 2 later in life, the cells would have been overwhelmed by another problem, such as obesity. Neither seems likely in Derek's case. The only other possibility is some sort of massive trauma to the pancreas. A gunshot, for example, a knife wound... or, theoretically because there's very little research in this area, a massive dose of radiation. A spike in diabetes cases has been noted in Chernobyl survivors. Radiation has been found to cause diabetes in animal experiments. Radiation is known to cause random changes to cellular DNA. If Derek had radiation in his pancreas, it's possible his insulin-producing cells malfunctioned. That could have led, according to Prof Kar, to high glucose levels, coma, and death. But it's all speculation, without a post mortem.'

She asked the MoD where Derek's post mortem was, without success. But without one, no doctor should have confirmed diabetes as the cause of death. There were three possibilities: either the post mortem existed but was being covered up as potential evidence of damage from radiation, or there was no post mortem and the cause of death was false, or he'd died from something else entirely. Either way, the *Mirror* felt, the family had been lied to, Parliament had been misled, and it stank.

By now, a former Army commando called Johnny Mercer was Veterans Minister. The role had often been occupied by an MP with a forces background, but Mercer was also young, media-friendly, and inclined to speak his mind. As Labour began making inroads, he agreed to meet campaigners to discuss a children's health study. Researchers confirmed it would be possible to simply extend the NRPB studies, but again it somehow never happened.

In December 2020, Winstanley wrote to the BNTVA with a medical decision. He acknowledged Williamson had supported it, but that wasn't enough. He wrote: 'In considering whether the type of service undertaken by personnel involved in the test programme could be regarded as medal-worthy, the overwhelming view was that, notwithstanding any findings over health risks, service in an austere environment did not, of itself, amount to the rigour deserving of a medal.'[26] The veterans were aghast - the nuclear weapons programme had been more than 'austere'. Owen, who had left the BNTVA and formed a charitable company called Legacy of the Atomic Bomb Recognition for Atomic Test Survivors, or LABRATS, said: 'The MoD now gives out medals to drone operators, sitting thousands of miles away from danger, operating a joystick.'

The *Mirror* published the revelation under the headline 'Betrayed By Penpushers', with the 'men from the ministry' once again blamed for denying justice to the veterans. Labour leader Keir Starmer - a centrist who took over from Corbyn after an election defeat - said: 'These servicemen were placed in huge danger, and many have suffered throughout their lives. The Prime Minister should personally intervene to ensure these heroes are recognised. This is not just a political, but a moral test of his character.' Hayes added a Tory voice of outrage: 'This decision flies in the face of reason. The veterans were in immense danger serving Queen and country. Leaving aside their health conditions, the massive personal risk they took warrants recognition.' Government minister Mercer, however, wrote to campaigners: 'Whilst disappointed with their decision, I understand it.' Shirley, as ever, had the last word. 'We fight on. We always have, always will. It's about time they realised that, and accepted the thirst for justice does not die when we do.'

Six days later, it was announced the decision would be reviewed. But while the delays dragged on, veterans were dying. The *Mirror* reported in December 2020 how Ken Miller, who believed he'd never had children because of his service on HMS *Warrior* at Grapple X in 1957, had passed away with no family to mourn him. The 80-year-old, who'd been married three times, was admitted to hospital with a chest infection and died soon after. His friend David Singletary said: 'He was sterile, had a double hip replacement, lost a lung. You're not going to tell me those things happen to every 80-year-old.' A call went out to find people to attend the funeral, which would otherwise have been witnessed by a lone council official. The BNTVA sent a coffin drape while forces charity SSAFA arranged a bugler, veteran pallbearers, and a packed crematorium. That week, Prime Minister Boris Johnson came under more pressure to overturn the medal decision. Ten of his own MPs, led by Hayes, wrote to him to press the case. French president Emmanuel Macron also piled in, announcing in February 2021 a medal for the veterans of their own tests in Algeria and the Pacific. It was on several hundred chests by that July. Britain's failure to do the same was becoming an international embarrassment.

In March, Shirley Denson was sent home to die. The chesty cough which had been bothering her for months turned out to be stage 4 lung and bone cancer, and soon after diagnosis she was receiving palliative care at home. One of the last things she did for Eric was to apply for the Elizabeth Cross in his name. Equivalent to the US Purple Star, it is awarded to the widows of those who died for their country.

With a war pension which accepted Eric had died as a result of his service, Shirley asked for the highest honour the state can give one of its troops. She said at the time: 'My Eric was ordered to put his life on the line for the sake of his young Queen and country. He did it without a moment's hesitation... he never saw his daughters grow up and have families of their own. I trust our Queen now, as I did then. I know she will see that justice is finally done.'

She was supported with letters from Hayes and Thornberry, who stressed the urgency of the case, but again the officials went slowly. Al Murray took on the job of emailing and calling the senior MoD officers whose job it was to decide on Eric's cross, telling them they had only days in which to make a decision. He told the *Mirror*: 'To wait a day longer, considering Shirley's diagnosis, would be not only immoral and unjust but a gross insult to her and Eric's service. When we applied for the medal we stressed Shirley's extreme age and the decades of injustice Eric's family has suffered... Five months is more than long enough to check his files.' Thornberry must have known she was asking for the impossible, but added: 'I hope - having heard about Shirley's illness - common sense, compassion and plain decency will prevail at the MoD.' Two weeks later, in April 2021, the cross was turned down on the basis of 'insufficient evidence'. Rear Admiral Jim Macleod said that while the pension was awarded on the basis of probability, medals required proof beyond all reasonable doubt, the criminal standard. Because the MoD doubted, therefore, there would be no medal.

No such rule had been included when the Queen constituted the gong. All other medals are given to those who are confirmed as taking part in a campaign, or who have documentary evidence of the risks they took for their country. Eric qualified on all counts. Murray asked for a second opinion but the decision was upheld just a few hours before Shirley died. Boniface did not tell the family until afterwards. Murray summed up the official attitude as 'bitterly disappointing but grimly consistent'. Defence Secretary Ben Wallace sent his condolences, and said: 'We gave very careful consideration to her application for the Elizabeth Cross but concluded the circumstances of her husband's sad death did not meet the eligibility criteria. The department does not in any way undervalue Sqn Ldr Denson's service. We are grateful to all those who participated in the nuclear testing programme.'

Boniface added her own comment: 'Shirley Denson had one word to describe the MoD and its 70 years of lies, denial, omissions and evasion. "Bastards," she said, every time I interviewed her. "Utter bastards." And when she died, they proved it again.'

Shirley was the first widow the BNTVA knew of to win a war pension. She was the first who'd got it for someone who'd been mentally ill. She uncovered the list of radiation dose badges kept by the AWE, and made it possible for other RAF veterans and families to make a claim. She coaxed and comforted other widows, and was one of those who took the MoD all the way to the Supreme Court. Her own family suffered as she devoted herself to campaigning, and romanticised a husband to whom marriage had, in truth, been very difficult. But having made it to 87, and the matriarch of a massive if troubled family, she had achieved an incredible amount. In an obituary for the magazine for tenants of Haig Housing, the military charity which had given the Densons a home after Eric's death, Boniface wrote: 'Right to the end, Shirley was in love with "my Eric". Whenever I saw her, she had a floaty scarf wrapped around her neck, forever the young woman who had just stepped out of the squadron leader's sports car. She was more than just a news story - she was the stuff of legends.'

Around the same time, Sandie Hern began chemotherapy for a cancer of her own. As she was nursed at home by Doug, the BNTVA to which they had both devoted their lives began to implode.

Owen's departure had heralded a change of trustees, and Doug was the last of the old guard to hang on. As historian, he had amassed huge piles of paperwork, and had no confidence others would look after it. The new team felt they had a right to it, but the dispute should have been easily resolved. Instead it led to a damaging sectarian war led by the new BNTVA chairman Ceri McDade.

Her RAF father Mike Marsh had served at Buffalo and Antler, and later suffered a duodenal ulcer and adrenal cancer which moved into his throat and pancreas. After she took over the reins of the BNTVA, she demanded Doug hand over his paperwork, and he resisted. After Sandie was diagnosed, he asked for time and space in which to care for her. One day, he came home from the shops to find Sandie collapsed on the floor in distress, having just answered the door to two police officers investigating a report of theft by the BNTVA. Hayes, as Doug's local MP, spoke to the police on his behalf. They agreed it was a civil matter, and told the charity there was no crime.

In January 2021, Boniface contacted both sides with a possible solution. The Imperial War Museum in London might take on the archive, and could protect and display it in perpetuity. She offered to put them in touch with the curator of their Cold War section, who might take it if both parties agreed. Doug opted to hear him out, but McDade refused. It was followed by an exchange of emails in which trustees accused Boniface of interfering and questioned her motives, and relations between the charity and the newspaper collapsed. The chance for a permanent exhibition at the IWM evaporated. In December 2021, the BNTVA lawyers wrote to Doug informing him they were going to sue.

McDade accused Boniface of 'mischievous intent and falsehoods' and 'bullying and meddling in the affairs of the BNTVA'. She wrote to the editor demanding 'an alternative *Mirror* journalist' work on the story, and the request was rejected. The acrimony led to the *Mirror's* stories being blocked or deleted from BNTVA social media accounts, and complaints to the press regulator IPSO about the accuracy of its reporting. None were upheld. McDade then wrote on behalf of the BNTVA to the PM saying the charity did not think it 'appropriate' to lobby him over the medal.

McDade had stepped down as chairman and taken up a salaried post as chief executive, but remained in day-to-day control. Doug appeared in court repeatedly over Zoom, with help from friends to operate the technology. Sandie died four months after legal proceedings began, and Doug became suicidal as the case dragged on. In January 2023, McDade was fired for gross misconduct. The BNTVA claimed to have found a catalogue of 'misdemeanours', and a £100,000 grant was withdrawn.

Doug's case was resolved six months later, when he handed over all remaining documents, and the BNTVA agreed not to pursue him for costs. It had spent more than £17,000, and the 'archive' turned out to include VHS video tapes of old reunions, Christmas card lists from the 1980s, and boxes of Minutes of the association. Much of the MoD and AWE documents in it had already been published widely, and it had never been worth the fuss.

Doug died three months later in September 2023, aged 86, before he could receive the medal he'd stormed Parliament for. At the reunion that year, Owen read out an apology from the remaining BNTVA trustees for the distress caused. 'We confirm that no theft of material had occurred, and that the matter had not been sanctioned by the board,' it said. After a short period of being 'technically insolvent', the BNTVA, which had done so much for so long, is effectively dormant.

Before Doug fell ill, he had secretly given the medal campaign a boost. In April 2021, 7.3m viewers tuned in to the BBC's flagship Sunday night show, *Call The Midwife*, as the dramatised story of a nuclear veteran, his young wife, and a deformed stillbirth was played out in primetime. The story of Derek and Audrey Fleming highlighted not only what happened to these young, fit men decades earlier, but the injustice and denials they endured. After their son's death, the MoD was shown refusing to share military medical records with civilian doctors. Doug served as a consultant on the show, and scriptwriter Heidi Thomas said: 'I was reading things from the *Mirror* campaign, and Dougie was a huge help. He changed a few lines, for example that the MoD had never fully admitted what had gone on. His was a real voice, coming from a real place, and so it was like having thousands of veterans looking at my script.'

The show's star Stephen McGann had been at the Hillsborough football tragedy in 1989, where 97 fans died after a crush and police covered up their culpability. 'I've seen what happened when smaller lives get shattered by a big event,' he said. 'The stories don't get told, and some get hidden when they should get talked about. We need to change the tone of how we talk about the nuclear tests, before these people die without the justice they deserve... those people mattered then, and they matter now.'

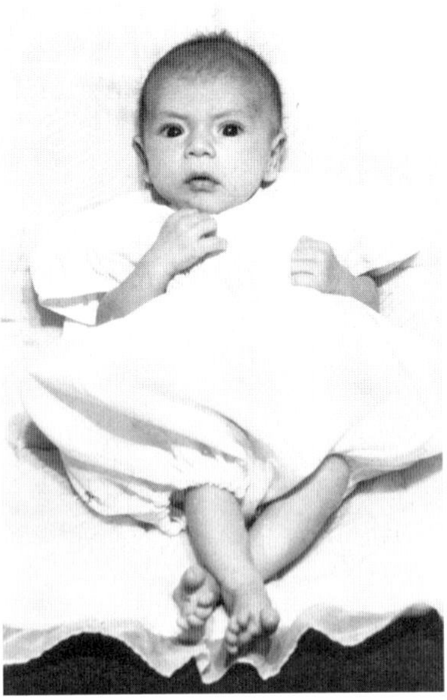

Steven Morris, pictured a few weeks before his sudden death at the age of just four months in 1962.

The broadcast provoked an outpouring of public support. A few days later, the *Mirror* published the very similar story of John Morris and his four-month-old son Steven. John had been sent to Christmas Island in August 1956 as an 18-year-old on National Service with the Royal Army Ordnance Corps. He worked in the laundry, and helped set up the atoll for the tests which were to begin the following year. He witnessed the first three Grapple atom bombs and Grapple X before returning home in November 1957. He met and married wife Betty, and she gave birth to Steven in late 1961. On the evening of Valentine's Day, they put the baby to bed as healthy and bonny as usual. But when Betty awoke and looked in his cot, he wasn't breathing. John awoke to the sound of his wife's screams, and began resuscitation while Betty pulled on her dressing gown and pelted out of the house to the nearest phone box. An ambulance arrived, and John recalled: 'We rushed him to hospital and the police were there. They took us off in separate cars, accusing us of killing him, smothering him or something.'

The couple were held separately in cells for most of the day, and it was mercifully soon apparent they were loving and devastated parents. They were released without charge, and went home in a daze. 'You have to face your neighbours, your friends, the pub, all these people asking what happened. And an empty cot,' said John. A post mortem found Steven had died from bronchial pneumonia. It should cause a fever, appetite loss, and breathing problems, but Steven showed none of those symptoms. The inquest noted that Betty told police her baby had been breathing normally. The pathologist who examined Steven's lungs found them covered in haemorrhages, which can be caused by an infection, but also by malformation of the lung's surface. Today, there would be a swab performed to confirm the presence of bacteria. In 1962, the only test was the doctor's best guess.

John never accepted the post mortem results, and spent decades trying to get a copy of the report. He said: 'Bronchopneumonia is about as likely as me climbing Everest.' Within two years of Steven's death, John was diagnosed with pernicious anaemia, which is caused by a failure of the bone marrow to produce red blood cells. It's such a well-known side-effect of radiation that cancer patients are warned they are at risk of it if they have radiotherapy. He has to have vitamin B12 injections every three months to counter it, but a war pension was rejected for lack of any clear proof of his dose.

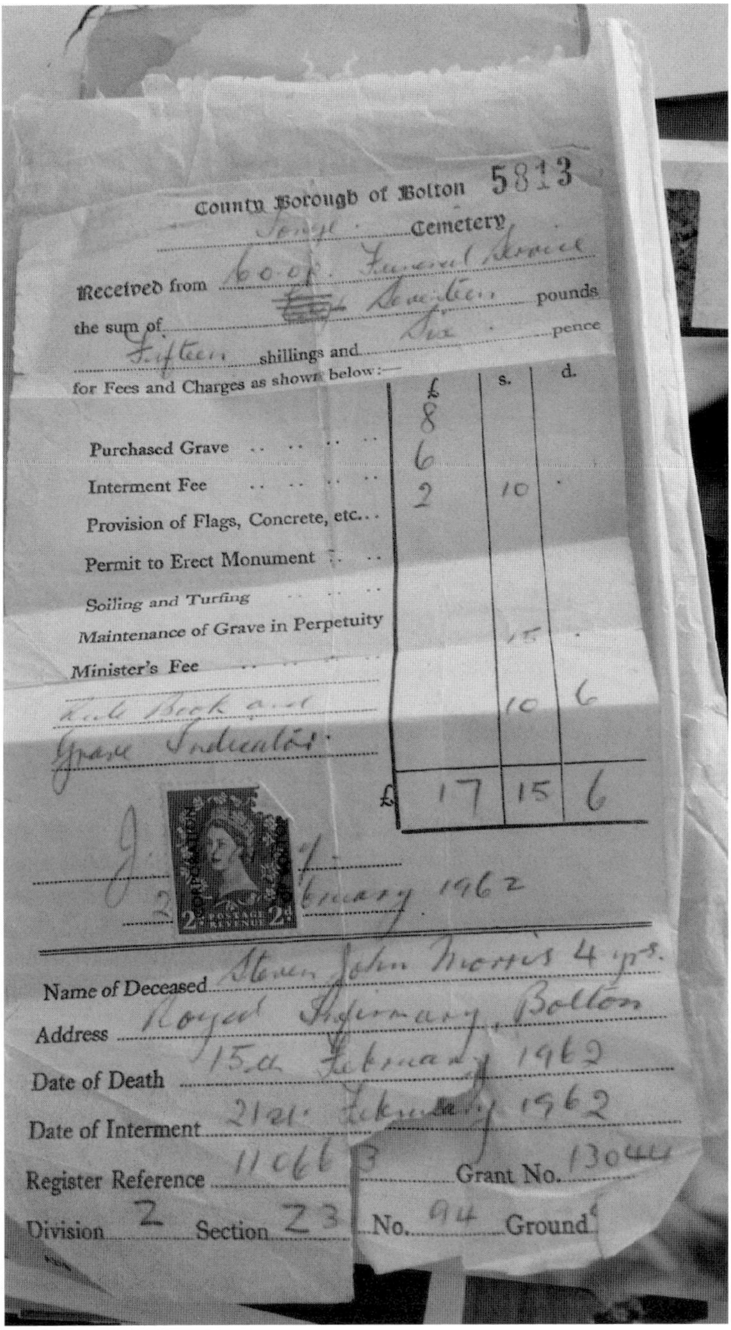

County Borough of Bolton 5813

Tonge Cemetery

Received from Co-op. Funeral Service

the sum of Seventeen pounds

Fifteen shillings and Six pence

for Fees and Charges as shown below:—

	£	s.	d.
	8		
Purchased Grave	6		
Interment Fee	2	10	
Provision of Flags, Concrete, etc.			
Permit to Erect Monument			
Soiling and Turfing			
Maintenance of Grave in Perpetuity		15	
Minister's Fee			
Rule Book and Grave Indicator		10	6
£	17	15	6

2 February 1962

Name of Deceased Steven John Morris 4 yrs.

Address Royal Infirmary, Bolton

Date of Death 15th February 1962

Date of Interment 21st February 1962

Register Reference 11066 B Grant No. 13044

Division 2 Section 23 No. 94 Ground

The receipt for Steven's funeral, which cost £17, 15 shillings and sixpence. His grieving parents John and Betty could not afford to pay, and neighbours held a whip-round to help.

Betty died in 2003, but they both struggled to come to terms with Steven's loss. They were terrified to leave each of their subsequent three children alone at night, so John changed jobs and took a pay cut so he could do shifts, and keep an eye on them as Betty slept. His children and grandchildren were all worried about having families of their own. And John still keeps the receipt for £17, 15 shillings and sixpence, which was the cost of Steven's funeral. 'It's a cover-up. I've had cancer, I've got a blood disorder, I've lost a child. But I'm one of the lucky ones. I woke up this morning,' said John. 'A bit of bloody metal is the least they could do. We deserve to be shown the country cares.'

Parliament held a debate about the medal. 'Time and again we have been blocked by a combination of the top brass... and the military establishment in the MoD,' Hayes told the Commons. 'The medal is a bang to rights case. The fact that this committee suggested an absence of risk and rigour is extraordinary. There can be no greater risk than going into a radiation cloud.' Mercer was the minister who responded for the government. He said the medal system was outside his control to maintain its 'integrity'. But he added: 'I am sure we will get there in the end. We will arrive at the right answer. And I urge [Sir John] to keep going.' Half an hour later, he was sacked.

It was time to think again, and the campaigners prepared another heave. Boniface and Owen formed a 'steering group' with John Morris, his granddaughter Laura, her MP Rebecca Long-Bailey and Steve Purse, to brainstorm ideas. On the Grapple Y anniversary, the *Mirror* challenged Prime Minister Boris Johnson to 'Look Them In The Eye And Tell Them Why They Don't Deserve A Medal'. Accompanied by a social media campaign of short videos, everyone from Keir Starmer to satirist Rory Bremner joined veterans and families to challenge the PM to 'look me in the eye', and explain.

There was John Sankus, who developed crumbling bones and spondylitis after Grapple Y. His wife had a miscarriage, a son born with encephalitis, a second with a dysfunctional thyroid, anaemia, crumbling bones, stomach and nerve issues and a nipple-like growth on his calf, a grandson with bone spurs and a granddaughter with a deformed thumb. It also featured the story of Tony Butler, an RAF corporal who had decades of skin cancers and teeth problems before an MRI discovered he was 'riddled' with tumours in almost every organ. He lost three stone in a month and died aged 72, in 2009. Three children born after he came back from the tests have auto-immune conditions. The paper listed the health problems of his youngest, Sarah: 'Postural orthostatic tachycardia syndrome, extreme fatigue, cardiac arrhythmia, vitiligo, eyesight problems, connective tissue problems causing her joints to dislocate and... atrial fibrillation in her teens. She had six miscarriages and her son has autism.'

Within weeks Starmer showed the PM the sort of headlines he was missing. The Labour leader invited the steering group to his Westminster office, and when they walked in the room, before anyone spoke, told them: 'I'm sorry for what you've been put through - not just the testing, but the absolute failure to support, recognise or address what came afterwards... The terrible decision to not award a medal was politically wrong, and morally wrong. The PM must now meet you, hear what you have to say, and do what it takes. We will support you all the way.'

He listened to the stories of Owen, Morris, and Purse, and the *Mirror* reported he was 'visibly moved' when hearing how Steve fretted every time his son cried. 'When I was a baby I screamed like crazy because I had hydrocephaly, so every time [he] cries I worry if he's developing it,' Steve told him. The Labour team in the room - Starmer, his shadow Defence Secretary John Healey, shadow Veterans Minister Stephanie Peacock and Long-Bailey - were reminded how the party had supported veterans before in Opposition, but turned its back in government. Twenty four years later, Boniface asked Starmer directly: 'Are you going to do the same?' He replied: 'No. The country owes you a huge debt. Your campaign is our campaign. It starts now.' Pressed on whether he'd stand by Corbyn's manifesto commitment of £50,000 a head compensation, he added: 'That's the starting point. We must do more. The next election is two or three years off and something must be done before then.' As the campaigners left, Starmer took Steve aside and told him how, as a father himself, he understood the worries that plagued him.

Within a week of the meeting, Labour was pressing in Parliament for the PM to do the same. Veterans Minister Leo Docherty replied that the medals system was 'an independent process and therefore not for ministerial intervention'. Then the PM was ambushed in Parliament. Long-Bailey challenged him in the Commons, saying: 'My constituent Laura told me Remembrance Sunday hurts. It hurts because there have been no medals for her grandad's service and the thousands of men involved in nuclear weapons tests between 1952 and 1991. It hurts because studies of such veterans have shown increased miscarriage, increased birth defects, and the same rate of genetic damage as clean-up workers at Chernobyl. And it hurts because the UK is the only nuclear power on Earth which has denied recognition. So I ask the Prime Minister, will he recognise nuclear testing veterans and agree to meet them? The Leader of the Opposition has.'

Johnson, who had no other option, replied: 'I will certainly make sure we get a proper meeting with representatives of the nuclear veterans.' But it would be another nine months before No10 made any effort to arrange it.

While they waited, the headlines continued. Northern metro mayors Steve Rotheram and Andy Burnham met veterans and told the *Mirror* they'd been 'victims of a crime'. The duo, who were campaigning MPs during the Blair years and big beasts of the Labour Party, compared the nuclear tests for the first time to other national scandals like Hillsborough, and the Grenfell Tower fire. 'The pattern of these scandals is always the same. They deny, suppress, cover up,' said Rotheram. Burnham added: 'These are the tactics of the British state: to deflect on to the victims, use a lack of progress to grind people down, and create mental torture so people cannot fight injustice.'[27] When 2022 dawned, it was now-or-never for the medal campaign. That October would mark 70 years since Operation Hurricane, and as the country prepared to celebrate the Queen's Platinum Jubilee with street parties, the *Mirror* demanded it also commemorate the Plutonium Jubilee - with full honours.

Call The Midwife revisited the story, following the Flemings through a new pregnancy and a nascent effort to catalogue the health problems suffered by veterans. In February, LABRATS held a seminar where veterans gathered to tell their stories. Burnham addressed the meeting, saying: 'This is the greatest injustice of them all, because it betrayed brave people who signed up to serve our country, and it inflicted an ongoing and repeated harm to generations. In this 70th anniversary year, the Prime Minister needs to make a national apology to each and every one of you, and every member of your families.'

He received a round of applause, but it was possible to hear a pin drop a few moments later when seven veterans gathered on the stage to share their experiences. Interviewing them over a video link, Boniface asked them to raise their hands if they had been diagnosed with cancer. All seven did so. Then she asked them to raise their hands again if they'd lost a child. All but one put a hand up. No-one who witnessed it was unmoved.[28] Even Boniface wiped away a tear, and she'd heard it all before.

By now Downing Street was in a muddle. Johnson was beleaguered by allegations of Covid lockdown parties, and his government was on the ropes. The *Mirror* reported how, one lunchtime, the PM's spokesman brushed off Labour's calls for compensation and a gong, saying: 'When it comes to medals, that's a matter for the AMSC... I'm not aware of any plans along [compensation] lines.' But three hours later, Defence Secretary Ben Wallace got to his feet in Parliament and announced another review, saying: 'I absolutely recognise that we are now the last country to [recognise them]... I've asked officials to go back and look at that again.'

The medal committee seemed to be the stumbling block, so it became a target. In February 2022, the *Mirror* reported: 'Establishment figures who refused a medal to Britain's nuclear test heroes have been showered with honours themselves - for overseeing Whitehall budgets, running sports events, and reintegrating the Taliban into Afghan society.' The panel included a NATO advisor, MoD accountant, ex-Royal Marine and a charity boss. It refused to publish Minutes, and held just seven meetings in 14 months, only one of which had full attendance. 'It sought no eyewitness evidence, did not comment on documents veterans said prove the risks they were exposed to, and then ruled service at the nuclear tests was... not dangerous enough for a medal,' said the paper. Colin Moir was scathing, accusing the committee of holding nuclear veterans to a higher standard while they were rewarded for 'driving a desk'.

Within days, the UK Health and Security Agency published the results of the fourth government study, ordered by Gavin Williamson four years earlier. For the first time, it found that nuclear test veterans were more likely to die, and more likely to get cancer than other veterans. Although the overall risk was small, at 2 per cent, the granular detail was devastating. The rate of chronic myeloid leukaemia - a disease caused by genetic mutations in the bone marrow, and said in the report to be 'radiation-inducible' - was 377 per cent higher than for other veterans. The risk of suicide was raised by 17 per cent across all the tests. People who served at Maralinga were twice as likely to get cancer of the brain and central nervous system. The scientists of AWRE had a 74 per cent increased risk of the liver cancer which had killed William Penney.

With the perspective of a longer time frame, the researchers could break down deaths and cancers by decade, service, and test. More than 11,900 veteran deaths had been traced, 4,118 of them as a result of a tumour. There were more than 50 pages of data tables. One showed that in the first 20 years, veterans were four times more likely to have liver cancer, 18 times more likely to get blood cancer, and were eight times more likely to get leukaemia. Even half a century later, those risks were still elevated among survivors.

On HMS *Diana*, 58 per cent of the crew had definitely died, more than half with tumours of some sort. From Buffalo, 85 per cent of the men were dead, a third from cancer, and with double the expected rate of leukaemia. Air crew from the sniff planes were four times more likely to have died from 'benign and unspecified' tumours, while the decontamination parties suffered five times as much leukaemia as they should have.

Each test seemed to have specific symptoms - at Hurricane the biggest increased risk was for death from melanoma, while at Totem it was prostate cancer. At Mosaic it was suicide, at Buffalo liver cancer. Antler's biggest risk increase was in benign tumours of the brain and nervous system, while at the Minor Trials the risk of almost every type of cancer was raised. At Grapple 123 and XYZ, and at Dominic, it was the same story - liver, or brain, or blood cancer risk was raised. In one data table it revealed that of *Diana*'s 282 crew only one, remarkably, had a dose recorded, while every officer in the Buffalo indoctrinee force, and every scientist did. Almost all the RAF crews, in the air and on the ground, who had flown in or handled the sampling planes recorded a dose as well.

The research was not perfect. It could only count the veterans who had died, or had cancers diagnosed, in the UK. A cancer registry wasn't introduced until the 1970s and earlier diagnoses were less reliable, so Billy Morris wouldn't feature, while a veteran was only considered to have died if their status on the national databases had been changed to reflect it, which meant émigrés like Joe Pasquini would have nothing registered. If their NHS data was inactive, the study couldn't pick it up, creating a potential error of apparently not-dead, not-cancerous veterans who could easily have died from cancer somewhere else. But for anyone who read the detail, the study still showed troubling results. The authors said that, if the doses on record were right, these patterns of cancer couldn't be connected to radiation. The obvious counter was that, if the doses were wrong, it would explain everything. Researchers didn't discuss the flaws in the dose badges or reporting systems, but did say that only 23 per cent of those in the study had a badge, and two thirds had a nil reading.

'The current analysis indicates that the possibility that test participation has caused a small increased risk of leukaemia... cannot be ruled out,' they wrote. 'Prostate, stomach and bladder cancer were higher... for benign brain and CNS cancers, incidence among the participants was raised in both the earlier and later follow-up periods compared to the national population... [the prostate cancer data] could indicate a detriment associated with the tests.' They ruled out smoking, or asbestos, or other chemical contaminants creating the patterns.[29] As the *Mirror* reported: 'For the first time it said [the veterans] were right. They are more likely to die and get cancer. Seven decades after developing a weapon always intended to be physically and genetically devastating to any living thing exposed to it, government scientists admitted that a "long-term detrimental health effect associated with participation in the tests cannot be completely ruled out".'

The MoD needed a moment. Soon after publication, it provided a holding statement, saying: 'This report is extremely important and we are studying its new findings closely... we will publish a comprehensive response in due course'. A day later, it had found some wriggle room, and said: 'As this latest report states, those who participated in the tests continue to show lower levels of mortality overall than expected from national rates, although this difference has narrowed with longer follow-up.' It ignored the fact that the study compared test veterans with other veterans, who are healthier than the general population, and on that metric those at the nuclear tests were more likely to get cancer and die. The MoD never did publish a 'comprehensive' response. More than a year later, Parliament was told in a written answer by junior Defence Minister Andrew Murrison that 'the results of these studies have consistently demonstrated that cancer and mortality rates for the nuclear test veterans are similar to those serving contemporaneously in the UK armed forces who did not participate in the testing programme, and lower than for the general population.'[30] It was a plain untruth as far as the first and fourth studies were concerned, and misleading. That was the MoD line, and it was sticking to it. But no-one, it seemed, was checking the footnotes.

The study was published alongside a 64-page 'companion report' with extra details. It said that RAF ground crew who washed down planes that had flown through the mushroom clouds were 5 times more likely to die from leukaemia. And suicide rates among test veterans, compared to other veterans, were excessive in every single decade since the weapons programme began: a damning indictment of the pressure they had been put under, during and after the testing. This was the 'extra detail' promised by Williamson when he ordered the study, and the report authors claimed the extra 19 years since the previous report made 'it possible to study in more detail and with greater statistical certainty the potential lifelong health impact of test participation'. Yet there were comparison tables, setting all this extra information against data from before 1998. And in those studies, there *was* no such detail. How, then, could it be compared? Owen asked the lead researcher where it might be, and was told that for the first time all four reports, or summaries, had been published together by the government on a single webpage, and each was accompanied by a 'companion report' with extra data.

The 'companion' for the first study in 1988 was 140 pages long, and told a very different story to the MoD. Just like the latest update, it had background tables that showed the risk of leukaemia wasn't the 'slight increase' long claimed by the MoD, but 350 per cent higher. Test veterans were 2.8 times more likely to die from bladder cancer, with raised rates of suicide. It included a table showing five test veterans dead from chronic myeloid leukaemia, compared to none in the control group, a statistically-significant figure with a low rate of probability, which meant that in 1988 there was proof men at the nuclear tests were 134 times more likely to get a 'radiation-inducible' blood cancer.

It mentioned the crew of HMS *Diana*, Buffalo indoctrinees and air crew as being more liable to exposure, even though the *Mirror* had first reported this as a revelation 20 years later, in 2008. It talked about the Blue Book which Shirley hadn't discovered until the mid-1990s, and which was trumpeted by the BNTVA at the time as groundbreaking. Something was off.

All the companion reports had ISBN numbers which traced back to The Stationery Office, the government's publishing arm. A check of the ISBNs proved the 1988 'companion' data had never been published, and not a single copy ever purchased.[31] All anyone had seen were the summaries. The 1988 one was published across seven pages in the *British Medical Journal*, and was the only version seen by veterans. Scanning it for any reference to this background data, Boniface finally found it - buried in a footnote. The text said: 'The description of the methods used to obtain accurate and unbiased information is necessarily long and is given in a report published by the NRPB.' In the small print, it mentioned a reference, NRPB-R214, and the ISBN number. It made no mention of extra data, which was odd for a scientific report, as if someone was trying to hide it. Simply put, the most shocking data had been suppressed for 34 years - hidden in a footnote with misleading language, deflecting the attention of all the journalists and campaigners who would have combed through it had they realised NRPB-R214 was not about statistical methodology.

Boniface rang McGinley, who had recently been diagnosed with Stage 4 lung cancer, and asked him what he remembered about the publication of the 1988 study. 'It was just a few pages in the BMJ, and it was a whitewash,' he said. But there were 140 pages of background data, had he ever seen those? No, he most definitely had not, and he was horrified. 'My God, how many men who got sick after serving their country were denied war pensions? Their wives denied the truth? And the MoD spent millions of taxpayer pounds fighting us, when they knew this all along,' he said. 'I remember [the revelation about how some groups were at more risk] being a massive admission from them in 2008, something we'd never heard before.' The *Mirror* said it had 'uncovered proof that SEVEN governments knew servicemen were more likely to have been exposed to radiation, get cancer, and kill themselves, for 34 years - and never published it.'

It was now four months since the PM had promised a meeting. Long-Bailey was called again at the weekly questions, and this time brought the House of Commons to a standstill. Raising the article published that day, she said: 'The Prime Minister has been very supportive of nuclear testing veterans, so I am sure that he will be shocked that, today, the *Mirror* has uncovered 140 pages of data previously hidden in the footnote of a 1988 government report. There is now concern that the High Court and this House may have been inadvertently misinformed in 2008, when told that only 159 men in UK nuclear weapons tests were exposed to dangerous radiation, when today's data shows exposure numbers were actually 2,314.'

She paused as an audible gasp filled the chamber, and MPs shook their heads. 'Will the Prime Minister urgently investigate this and arrange to meet personally in Downing Street with my constituent, her grandad and other nuclear testing veterans to bring an end to this national scandal?' Johnson, who had no alternative but to agree, replied: 'I thank the honourable member very much for bringing those facts – new facts – to the attention of the House, and I know that my office has already been in touch with the group concerned to make sure that we have a proper meeting. I hope very much that she will be there, and we will be able to discuss all the issues that she has raised.'

Left: Kenneth Measures, pictured in navy uniform before he was posted to Grapple where his job was to wash helicopters that had flown into Group Zero collecting samples. (Courtesy of Melanie Measures)

Below: Kenneth Measures in a photo he sent home, taken while he swam in the seas off Christmas Island. He died from rare radiogenic cancer 36 years later, but the coroner was told he was not exposed to any radiation. (Courtesy of Melanie Measures)

In the same year the NRPB study was set up, Kenneth Measures died, aged 54. His wife Valerie wrote in a letter to their daughter Melanie at the time: 'They give him all kinds of medicine but we know nothing is doing any good, for the cancer is spreading fast... his poor body is in such a mess. The lower area and legs are so swollen, but the top of him is like something out of Belsen. It breaks my heart to see it.' She had nursed him at home while she could, taking time off work, leaving early, and going in late. When his pain was too great he went into the local cottage hospital in Helston, Cornwall, to die. His post mortem shows by that point he had secondary tumours on his spine, and his body was terribly wasted. Ken had been an aircraft fitter, based on HMS *Warrior* and *Narvik*, for the first three Grapple tests, and Grapple X. It was his job to maintain and wash off the barrier paint from helicopters which had flown into Ground Zero.

He was always fighting fit, continuing to work on navy helicopters as a contractor in Civvy Street. Letters he wrote in 1982 to Melanie when she was living in Italy, gave her the local weather report, stories about the family cats, and begin complaining of a dry cough. It spreads into pain in one side, and when cancer was found he had a lung removed. His letters stay cheery, but then new complaints creep in - stomach pain, outpatient appointments, x-rays and scans. In the summer of 1983 the letters stop and Valerie takes over, warning their daughter not to ring the house, but to call her at work for updates about her dad, as she was trying to keep the truth from him. He had adenocarcinoma of the lung, a radiogenic cancer which is rarely seen in someone so young.

After his death, Valerie demanded a post mortem, and persuaded the coroner's officer to investigate his service at Christmas Island. An inquest was held, at which AWRE scientist Edward Fuller testified the helicopters were not contaminated, their samples were in 'impervious' polythene bags, and any fallout Ken had ingested would have 'decayed rapidly within the body'. In February 1984, a defence minister confirmed in a letter to Valerie's MP: 'The information from the AWRE shows Mr Measures received a zero radiation dose from the United Kingdom nuclear tests.'

The inquest reached a verdict of death from natural causes, but Valerie told her local paper: 'I will fight on, and investigate in my own quiet way. There must be something there... there was a lot not said, a lot that could not be said.' She told a BBC producer that the inquest was 'a farce from beginning to end, and I don't like being made a fool of'. She added: 'If one fights governments, one must have ammunition.' She died at the end of 2023, aged 93, suffering from dementia and no longer able to recall her futile battle for the truth. Melanie said: 'She just hit a brick wall, and in the end it made her mentally and physically ill... She got depression, and it was decided that it was better to just shelve it. She put all my dad's stuff in a box and put it away.'

Two months after her death, the *Mirror* found proof that the MoD had misled the coroner. The 140-pages from the R214 report had a crucial nugget of information: the dose badge data which had been relied on to deny any liability was bogus. The report states the badge data was collected by atomic scientists and summarised in the 1960s, and the badges destroyed. When it was shared 20 years later with the

NRPB, they were told the only notes made were from badges showing above the 'minimum recordable level' - which kept changing, and was not zero. By the end of the test programme in the 1960s, the minimum recordable level was 0.1mSv, but the normal figure in Australia was 0.2 mSv. At some operations, it was 0.3 mSV or 0.5 mSv, and at Operation Buffalo the minimum was 4 mSv. These amounts equate to anywhere between two weeks' and two years' worth of normal background radiation. Anyone with such a dose on their badges was noted down as 'nil', even if repeated badges showed the dose accumulating. The *Mirror* explained: 'A dental x-ray involves 0.005 mSv of radiation, and even that low dose can increase the risk of cancer... [at Buffalo] more than 200 officers could have been exposed to the equivalent of 800 x-rays, and would still be recorded as having zero radiation.'

The paper said the dose badge data was used to bat away inquests, pensions and court claims, and misleading evidence given under oath. MPs went on the record, demanding the badge data be reviewed or dumped, but when the MoD was asked to comment it insisted that its policies would not change.[32] Melanie said: 'What mattered to my mum is the way people were treated, not just with the lack of information but being brushed off like they were nobody. I just want the truth to come out. My dad was so proud of being in the navy, and serving the Queen. Yet when he needed help there was none.'

It was becoming clear there had been a concerted effort to obscure the truth. Scientists had falsely described data as methodology, the AWRE had stated under oath and to multiple judges that veterans had a zero dose knowing the badges were questionable, and despite growing evidence the MoD press office was claiming black was white. Either the MoD was engaged in a deliberate cover-up to protect itself from financial liability, or a bureaucratic institution was being complacent, unable to consider the possibility that it had ever gone awry. Possibly both at the same time.

That the machinery was malfunctioning - or at least finding the nuclear veterans hard to digest - became clear in late 2018. A new press officer had been deputed to deal with Boniface's latest request for comment, and rang to apologise it was taking several days. He told her: 'This one has to go through more levels of sign-off, with more people, in more offices, than any other story I've ever dealt with.' Boniface, who already knew the MoD usually answered in a single day but with test veterans habitually took three, laughed, and said: 'Has it got to go to the brigadier in the basement?' What was a joke became a conviction over subsequent dealings, as Boniface realised the press office responses were being drafted by government lawyers. The same line would be followed for months, on stories that should ordinarily require a fresh take. Any request to amend it, or answer a specific question, was rebuffed, not because the press officer did not want to, but because the lawyer would not let them. She later explained: 'It passes across some very shiny desks, and I've also found out my FOIs go across precisely the same desks, one way or another.' She imagined this fictional brigadier 'with a blank where a face should be, a regimental tie, and furiously shiny shoes. He sits at the back of war pension hearings, he floats in and out of public galleries, he is unremarkable... he is Agent

Smith, he is Randall Flagg, he is the Smoking Man. He is it; them; the man.' Aware she sounded mad, she clarified. 'He is the embodiment of an institutional inability to admit to errors, and to compound them by committing more... an idea that I have to find a way to beat. Every day I wake up and think, how can I piss off the brigadier in the basement today?'[33]

And then came evidence that the science had been manipulated, as McGinley had instinctively known decades earlier. Documents emerged showing that an MoD official asked the NRPB to rewrite the first study to be more favourable - and his demands seemed to have been met. That official was Mike McTaggart, who had previously pressured the Department of Health on pensions, conspired with the Australians to manipulate FOI requests, and gleefully noted 'we can forget it!' after Parliament ran out of time to discuss test veterans. In 1988 he wrote a letter complaining about an early draft of the study to NRPB director Roger Clarke: 'The clear implication is given that all participants have been caused harm.' He added that most people would read only the study's opening and conclusions, and 'it is therefore necessary to couch these sections in terms which... prevent the reader reaching a distorted view.'

He wanted to describe a 345% rise in the rate of deaths from radiogenic leukaemia as a 'slight increase' compared to national rates. He says it 'at best leaves the casual reader with the impression that the excess is real, it was caused by

Liverpool mayor Steve Rotheram waits his turn as Greater Manchester mayor Andy Burnham signs a banner promising their support for the medal campaign, with John Morris and Archie Hart looking on. (Susie Boniface)

radiation and that there is evidence of ingestion of long-lived radionuclides, and at worse gives the veterans a bonanza of phrases to quote out of context'. He adds the figures could not 'be lightly dismissed as a chance finding', but then criticised the draft for calling it a 'chance occurrence' only once. How the scientists responded to this highly unusual and unethical interference is unknown, but in the final version of their report, a tripling of leukaemia risk is called a 'slight increase' and the death rates are blamed on 'chance' 22 times. The *Mirror* asked the MoD to comment on claims the science had been discredited, and whether the data would be given to independent researchers to verify. Answer came there none.

But the veterans were no longer alone. Burnham and Rotheram asked them to join the Hillsborough Law campaign, which calls for legislation to prevent future cover-ups by giving public officials a legal duty of candour, and establishing a public advocate to represent victims without having to fund their own lawyers, which puts them at an instant disadvantage when tackling massive organisations lawyered up to the hilt. By uniting a dozen or more campaigns, the mayors' intention was to not only prove the need for systemic change, but give them all extra credibility and clout.

The two politicians had something else up their sleeves: experience. They talked at length with the campaigners about how to break a scandal open, and how they had done it before. As a health minister, Burnham had once dismissed the claims of patients to have been infected with contaminated blood products in the 1970s, but changed his mind after sitting down to hear from them direct. He was shown medical records claiming one of the victims was an alcoholic, which had caused his liver disease, when in truth he was a teetotaller who got hepatitis from a blood transfusion. 'Altering medical records is a crime,' he said. 'If you can prove they've done that, you can force the lid off this.' Boniface had long heard veterans claiming their records were missing; perhaps there was some way to prove it.

The immediate problem was still getting a medal, while there were enough survivors left alive to make it meaningful. Their average age was now 85, and despite Johnson's promise of a meeting it had yet to happen. Boniface had been emailing No10 advisers for weeks, and in February 2022 there was finally a response. It was not the Downing Street showdown that was hoped for, but a 'private meeting... while the PM is on a trip outside of London'. The idea was for Johnson to meet veterans after a visit to an RAF base in Lincolnshire - a rural spot hard for wheelchair users and the elderly to get to, and a no-go as far as the *Mirror* was concerned too. It wanted the veterans treated with honour at the heart of government, not a quick sit-down in a dingy back room after the PM had done a photo opp. It was refused, and it turned out to be the right call - Johnson was later compared to Kim Jong Un for posing in the cockpit of an RAF P-8A Poseidon jet, which had flown a round trip of 660 miles just for the photograph.[34]

When he went back to Christmas Island, Ray Carbery had said: 'They didn't send the weedy guys. They gave us all medicals first, we were the healthiest they had. The surprise isn't that some of us aren't so bad, it's that so many healthy men aren't here any more. Whatever happened on Christmas Island killed the fittest lads in the armed forces. Out of the 12 in my unit, half died of blood cancer.' After

Burnham said medical records might lift the lid, Boniface found in her files two pieces of paper about blood testing to be carried out on all visitors to Maralinga in 1957. Tom Watson, when he first became involved with the campaign, had asked a question about it in Parliament in 2018, and been told: 'The MoD is unable to locate any information that suggests that AWRE staff took blood samples for radiological monitoring at the tests. Service personnel who were present may have had blood samples taken during their career, but these individual military medical records are not held centrally.'[35] It had seemed a dead end.

Then a few weeks after coming back from Manchester, she got a WhatsApp message. Joe Pasquini's papers from decades of research had included a memo between AWRE scientists, discussing the 'gross irregularity' of blood counts taken from a pilot who had flown through multiple mushroom clouds. The sender had blurred the name for privacy reasons, but after reading countless documents about the tests Boniface had a clear idea of who it was. 'Good God. It's Gledhill isn't it?' she asked her source. Terry Gledhill, awarded the Air Force Cross for his bravery at Grapple, had been the squadron leader for the sniff planes. He made the first 'cuts' into the clouds, ruled it safe, then ordered his men to follow him in. On the memo were the results of seven blood tests taken between August 1957 and November 1958, and a scientist asking what 'further action' should be taken. The document had a National Archives reference, but while it had briefly been made public - when Pasquini took a copy - it was one of the ES files shut in the 'security review' in 2018. It was a piece of paper no-one was supposed to see, and it was dynamite.

Boniface had been sceptical of veterans' claims to be 'guinea pigs'. Negligence seemed more plausible than cover-up. But here was evidence she was wrong - they were experimented on, all along. Blood testing was the only way to find out if bone-seeking fission products like plutonium had got inside the body, and were having an impact on how the bone marrow produced new blood cells. That in turn could influence health, from the type of anaemia reported by John Morris to rare infections, or cancers. If they were testing, there was a risk. If the results were withheld, there was a lie. And if they had tested one pilot, they probably tested more - and perhaps that Maralinga document was not a dead end at all.

Squadron Leader Terry Gledhill won the Air Force Cross for his bravery leading his pilots into the mushroom clouds of Operation Grapple. After his death, his family continued his long fight for the truth. (Courtesy of the Gledhill family)

Gledhill had died in 2015, but daughter Jane O'Connor said he had long suffered unexplained illnesses, infections, fatigue, bruising, and never had any answers to his questions about whether Christmas Island caused it. She was astonished to see the memo, and readily agreed to make a formal request for her father's medical files, including any blood tests, to the AWE and RAF. The boffins sent her a copy of the same memo, but the MoD refused, point blank, to provide anything. The organisation which held medical records was officially withholding them from those who might have a legal claim against them. This was no longer an historic cover-up, but a current one - and someone working for the MoD today was possibly committing a criminal offence.

The memo linked Gledhill's testing to an Air Ministry order, and Boniface dug through the archives to find it. Sent to all RAF bases in April 1958, it was an instruction for any 'persons regularly exposed to radiation' to be given blood tests to monitor their health, with limits below which personnel must be pulled from duty. If Gledhill had qualified, then so would all aircrews at Grapple Y and Z.

This was devastating, but the campaigners decided to sit on it. Nothing could be allowed to derail the medal, while there was still a chance thousands of old soldiers could get it before they died.

When the medal committee was criticised, a government spokesman always insisted that its 'decisions are independent of government'. But evidence was found proving beyond all doubt that the MoD had tinkered with the process. The *Mirror*

Terry Gledhill, second from right, stands next to one of the 76 Squadron Canberra bombers after the end of Grapple Z - an operation he was not allowed to fly in, having 'exceeded his dose'. (Courtesy of the Gledhill family)

printed emails, gained after a six month FOI battle, which showed officials passing the committee chairman 'lines to take' from 'our compensation colleagues' - which could mean either government lawyers, or war pensions staff. Neither are supposed to be anywhere near medal decisions.

The paper said the committee had 'wrongly linked medals to the issue of whether servicemen were injured - in clear breach of the rules... emails show an unnamed civil servant telling committee chairman Charles Winstanley: "I thought you might be interested to see the lines to take that our compensation colleagues use in response to the nuclear test veterans' claims".' They were the same lawyer-approved lies pumped out by the press office - that the veterans' protection was a vital consideration, there was thorough radiobiological monitoring, and no evidence they were sicker than anyone else. The chairman had called the information 'succinct and relevant', yet had later written to the BNTVA, in a letter published on the government website, claiming health issues had made no difference. Laura Jackson pointed out the inevitable failure of logic: 'Saying these men were only at risk if there is proof they were exposed, is like saying soldiers are only at risk of being shot if they have a bullet wound.'

Greater public and media scrutiny was causing the MoD to stumble. Now it took a serious beating, this time from a cache of 1,000 historic documents that had been declassified for the Royal Commission, and later assessed by Rosenblatt's for its lawsuit. They were never used in court, and instead of being sent to the National Archives were squirrelled away at King's College London. This was the evidence that the Supreme Court had never seen, and it proved officials knew of the 'serious biological risks' they were ordering men to take.[36] The *Mirror* reported: 'The documents show that cancer-inducing radiation doses were labelled "safe", kit and equipment was irradiated, servicemen showed signs of radiation in their urine, and scientists complained about having to observe "an unnecessary margin of safety"... The unearthed papers show that "safety limits" involved significant amounts of exposure. A dose considered to be a "zero risk" was equivalent to 30 years' of background radiation in just 5 days. A "slight risk" was the same as that of 250 years' background. Today, scientists believe even this natural radiation may lead to thousands of deaths every year. The documents set a "normal working rate" greater than most Britons would receive annually, and set different decontamination standards for scientists and servicemen even though they worked side-by-side. Scientists knew at the time that normal background radiation can cause cancers, and that there is no safe dose... After the tests, traces of radiation were found in servicemen's urine, in human bones, and in the bones and organs of animals throughout the areas they lived and worked in.'

The papers discussed 'observable genetic effects' which would appear 40 years later, and that contaminated Land Rovers were cleaned with a standard vacuum cleaner, with the filter bags shaken and beaten by men to clear it of toxic dust in a way described, with typical British understatement, as 'somewhat hazardous'.[37] One document, about plans to gather evidence of ingested fallout in 1955, stated: 'Human beings in a contaminated zone could absorb fission products from the air they breathe, from water or from food onto which fission products had fallen, or from the produce of animals which have previously ingested fission products.' In other words, exactly what veterans claimed had happened to them, and which the MoD had always denied was even a possibility.[38]

Scientists had found tins, medical supplies, and soap had not only absorbed radiation, but the very metal was contaminated on an atomic level. McGinley told the paper: 'The animals were studied to find out how much radiation they had absorbed. We weren't. We were second-class guinea pigs.' Radiation epidemiologist Mark Little added: 'I am surprised to see even back in the 1950s a statement that a zero risk was equivalent to this low dose. Although the risk is lower, it is most certainly not zero.' He added that the 'zero dose' daily rate was equivalent to a single CT scan, and that studies showed repeated scans can triple the rate of brain cancer and leukaemia. And in direct contradiction to the government line that no-one at Christmas Island was exposed due to a lack of granite underfoot, one document stated bluntly that sorties towards Ground Zero had exposed some personnel to a dose of 50-60 Roentgens per hour - more radiation than anyone could expect to receive in two centuries in the UK.

A month later, there was more - 300 pages of memos about HMS *Diana*'s mission, revealing the aim was 'to gain first-hand experience about fallout and contamination' and hazards to personnel 'as a result of passage through fallout from atomic weapons'. It had the same, unsafe 'safety limits', and showed the Medical Director General of the armed forces approving their use because 'in these infrequent, and very expensive, operations it is reasonable that some slight risks should be incurred by personnel rather than that important records and observations should be lost'. The risks were far from slight - the papers showed loose contamination throughout *Diana*, on decks, ladders and vents, and able to be easily transferred to lungs and stomach.[39]

Emails shuttled back and forth between Boniface, Long-Bailey, and No10, with Hayes whispering in ears in the Commons and the *Mirror* harrumphing in headlines. Finally it was agreed there would be two meetings - the first in the PM's Parliamentary office to hear the evidence, and a second in Downing Street where he would deliver a verdict. In essence, the PM had already decided to give the medal, and was doubling his PR. The fact his government was in its dying days, with constant badgering over broken promises and party failures, as well as a Parliamentary ethics investigation, made it all the more imperative he get some good headlines.

Alan Owen's heart burst on April 3. Having always felt he had avoided the legacy of the tests that had claimed his father and brother's lives, and caused his sister such pain, he did not expect radiation to catch up with him. During a walking football match at his local leisure centre, he suffered a cardiac arrest. 'I was dead on the floor,' he said. Luckily his teammates included ex-soldiers, there was a defibrillator nearby, and they kept him alive long enough to get him to hospital.

Fitted with stents and an internal defibrillator that would jumpstart his heart if it happened again, he explained the family history to the doctor, who immediately called in Alan's teenaged son for genetic screening. His condition is known to be caused by radiation, which can trigger the walls of the artery to thicken and create blockages. The blood backs up, the heart stops, and sometimes the artery bursts, which is rarely survivable. It happened to Alan's dad three times, and his brother and his sister just once. He began a long process of rehabilitation, but became more determined than ever to win justice for the nuclear veterans and their families. 'I am still here. Still campaigning and still pushing for recognition,' he said. Nine weeks later, a little grey in the face, he walked into Parliament and told the PM he'd cheated death to be there.

For weeks he, John Morris and his granddaughter Laura, Steve Purse, and Boniface had all rehearsed short, five-minute speeches that got their main points across. On the day, they marched together up to the Parliamentary security guards, along with Alan's sister Laura and Steve's mum Jacqui, both acting as carers. 'We're here to see the Prime Minister,' Boniface told them. 'He's expecting us.' Inside, they were escorted to the Speaker's Gallery to view the weekly bunfight of Prime Minister's Questions. Afterwards they met up with Sir John Hayes and Rebecca Long-Bailey, and were taken behind the Speaker's Chair to the PM's private office, overlooking the Thames and filled with officials, press officers, and a junior defence minister, as well as the PM. Boniface was first in the door, shook his hand and told him: 'After we leave, you'll be told these people are making it all up. But you're the only one who can fix this.' Johnson said nothing, but looked bemused.

Everyone sat around a large polished table, with John and the families facing the PM. John, as the only veteran in the room, spoke first and the carefully-scripted plan went out the window as he got into his stride, and some of the most powerful men in the kingdom sat in silence, taking notes, as he spoke eloquently about the bombs, about Steven, and his long battle for the truth. The *Mirror* later reported: 'John banged the table in the Prime Minister's office in Parliament as he said: "It's the ideal moment for you to look me in the eye and tell me you deserve a medal. Or say, sod off... We protected this country by providing the nuclear deterrent, but we've been let down by successive governments. My wife and I were arrested on suspicion of murder, and I've had a lifetime of worry and illness. I had blood and urine taken which is not in my medical records, and I've been refused a war pension." Holding back tears, John's voice broke as he added: "I just want justice for my family, and for Steven." The visibly-moved PM told him: "I'm so sorry, so sorry for your loss. You need a much better picture of what on Earth is going on".'

When Steve Purse discussed his dad's service at Maralinga, Johnson replied: 'Where? It's not something I've been educated about before... it's extraordinary.' When the campaigners mentioned their missing medical records, Boniface took a copy of the Gledhill memo and flicked it across the table at the PM, telling him that if blood tests were taken, and were now being withheld, it was potentially a criminal offence. Johnson agreed: 'Yes, it is.' Then he asked her where she thought they were. He said: 'There are things you are not being told and should be told. If it's been hidden away like in the Raiders of the Lost Ark, or stuff is being stashed in a vault by the British government, that needs to be sorted out.' Alan told of his scrape

Those who were in the 'Look Me In The Eye' meeting with the Prime Minister, pictured full of smiles afterwards. Behind, left to right, Laura Morris, John Morris, Laura Jackson, Alan Owen, Rebecca Long-Bailey and Jacqueline Purse. Susie Boniface stands behind Steve Purse. (Susie Boniface)

with death, Laura recounted her own troubles. And Boniface asked for a Plutonium Jubilee service of commemoration at Westminster Abbey. The campaigners left the meeting jubilant; and less than a month after he promised them justice, the PM was forced out of office. There would be no second meeting.

But if only a desperate PM would meet the veterans, then one with nothing left to lose might be able to deliver. In the month before he left Downing Street, the campaigners pulled out every stop to influence Johnson's exit in the veterans' favour - emailing, calling, and sending politicians of every party in to plead their case. Two days before his resignation the PM reappointed Mercer as Veterans Minister, and gave him Cabinet status, putting him at the top table of government. On the day that Johnson quit, Hayes went to see him to commiserate, and pressed him to deliver the medal before he departed. The PM promised to do just that, before turning to Mercer, who was also there, to say: 'Get it done.'[40]

Johnson's last act in office was to write a public letter to the test veterans. It contained the promise of another medal review, some funds to mark the Plutonium Jubilee, and the warmest words from a PM these men had ever heard.

'For decades our country has lived in peace and security under the protective shield of our independent nuclear deterrent. But we should always remember that we only possess this vital insurance because of your achievements,' he wrote. 'At a time when the essential technology was still in its infancy - and the world was even more dangerous than it is today - you succeeded in developing and testing our deterrent in the first place. You forged and perfected our country's protective shield and then made certain that it worked.' He added: 'Today 67 million people in the UK - and our NATO allies - live in the safety that you helped to provide. You can take pride in how our prosperity and security derives, in good measure, from your effort and sacrifice all those years ago.'

He said he was 'privileged' to meet them, and added: 'I'm determined that your achievements will never be forgotten... I have asked that we look again at the case for medallic recognition, because it is my firm belief that you all deserve such an honour, and this work is now in train... Many of us have spent our whole lives under the formidable shield that you helped to build. On behalf of all those millions, let me offer my profound thanks for your part in keeping us safe.'[41]

They were words dashed off by a PM desperate to be liked by someone, anyone, and who had lived by the fruits of his pen for decades before becoming party leader. They were also indisputably the most words any PM had ever devoted to these men, and at a time when Russia had just invaded Ukraine and a European land war was underway, what had happened so long ago seemed of pressing importance. Johnson had ensured that the crowing around his downfall was tempered by the reluctant praise of the *Mirror* - coincidentally, the same newspaper whose investigations into his wrongdoing had precipitated his end.

Attention now turned to his replacement. The two front-runners, Liz Truss and Rishi Sunak, both confirmed their support. Truss also vowed to look into war pensions, while a spokesman for Sunak said he 'would also back an investigation into whether the tests represented a criminal offence'. Truss won the party vote and took office, whereupon the Queen died, the country went into mourning, and Truss ignored the veterans completely.

Once the state funeral was over, it became obvious the radio silence wasn't just because the new PM was busy. She fired the Cabinet Secretary, who was also the chairman of the main honours committee, rendering it unable to rubber-stamp any medals. Then she fired Mercer.

The Labour Party conference that September seemed a crucial opportunity to not only register the government's failures, but to cement the Opposition firmly into support for the veterans. Polls were tipping consistently Labour's way, and it seemed likely they would form the next government, whenever an election was held.

John Morris spoke at a fringe event, organised to hammer home the message to activists and officials who form the party machinery, and was given a standing ovation by a room that did not have a dry eye in it. 'I watched the funeral of the Queen and there were people with medals because they'd been grouse-shooting on the moors. We gave this country its nuclear power and we have never been thanked or rewarded,' he said. 'I will fight the Labour Party if it does not honour its promise. I will be a thorn in their side. We're relying on you.'

Shadow Defence Secretary John Healey praised Long-Bailey for getting Johnson 'to do what he never does, and keep a promise'. He told Boniface: 'You with this campaign are taking on the whole British establishment.' And he told the audience: 'There is no good reason, no good moral reason, no good military reason, for withholding the recognition and compensation that other countries have had.

Shadow Defence Secretary John Healey at a 2022 Labour Party conference event, with Steve Purse, Rebecca Long-Bailey and, foreground, John Morris. Healey said 'there was no good reason' to withhold recognition or compensation from the veterans. (Susie Boniface)

Your campaign is our campaign. It is a pledge we are determined to deliver'. Andy Burnham added: 'There is far more than a medal needed here, it's truth, it's justice and reparation for the nuclear test veterans and their families. But it's more than that, we have to change the way this country works, and level up the scales of justice.'

A week later, Mercer wrote a joint letter with Labour MP and ex-Paratrooper Dan Jarvis, accusing the PM of 'abandoning our most mistreated heroes in her race for power'. The next day the *Mirror* claimed Truss had 'caved' after £450,000 was made available to mark the anniversary, to be spent on an oral history project and charity work. It would be coupled with a ceremony, not at Westminster Abbey as requested, but the National Memorial Arboretum in Staffordshire, home to hundreds of service memorials. But she ignored all requests to meet them, and there was no word on the medal. Mercer hit out at the delay, saying: 'The test veterans have been fighting for recognition for 70 years and for them to have to wait a second longer is simply unconscionable. They are in their 80s, and they are rightly losing patience. We know that previous decisions that went against them were held back, for months, as part of a Whitehall media strategy. These men delivered our nuclear deterrent, and they do not deserve to be used as pawns in cruel, political games.' Then came a week which changed everything.

The veterans protest at the Cenotaph in November 2022, having learned the medal committee had refused their request again. Holding the banner, left to right, are Jacqueline Purse, John Morris, Ernest Bow, Brian Unthank, Bill Lawrie, and Lucy Musselwhite. (Susie Boniface)

As Remembrance Sunday approached, the campaign found another gear. On the Monday, the *Mirror* had a leak confirming the medal committee had once again rejected the veterans. It reported: 'It comes despite pledges from the past three Prime Ministers to back one... A Cabinet Office source said of the snub by the Advisory Military Sub Committee, which refuses to publish its Minutes or verdict: "I don't know why they have come to that decision. It's a disgrace".' On Tuesday there was disdain from the MoD, after Long-Bailey asked the PM to overturn the ruling because the medal rules 'state clearly that the Prime Minister can personally make a direct recommendation to the sovereign on a medal issue'. Murrison told her snootily: 'She really ought not to believe everything she sees in the pages of the *Daily Mirror*.'[42]

That Wednesday was the 65th anniversary of Grapple X, and at a protest in Whitehall veterans wore a 'Missing Medal', an enamel pin badge the size and shape of a medal ribbon, in the danger colours of yellow and black. Archie Hart and John Morris, along with Steve Purse and other families, laid wreaths at the steps of Downing Street and the MoD, holding aloft a giant orange banner asking 'Would You Fly Through A Nuclear Cloud?' At the Cenotaph the old soldiers saluted, and Julie Soane wept as she laid a wreath for her dad, who was at Grapple Y. She said: 'He had bowel cancer and a brain tumour and I'm blind in my left eye. It means a lot to be able to lay this wreath.' At Downing Street the policemen watched and said nothing. But at the MoD, armed officers swarmed around the veterans as they observed a minute's silence, and later removed the wreaths.

Friday was Armistice Day - the anniversary marking the end of the First World War, when the entire British state focuses on honouring its veterans. Now the *Mirror* let rip. The splash headline was 'HEROES DENIED TRUTH', over a picture of Terry Gledhill's plane circling the mushroom cloud. Across five pages, it exposed for the first time how blood tests showed 'gross irregularities', and had been hidden ever since. 'Lies for 70 years,' bellowed the inside pages, as Labour figures called it a crime and demanded a public inquiry. Sunak, the new PM and 16th person in that job to have failed the nuclear veterans, refused to comment. After all, what could he possibly say?

The culmination of the week's publicity was the Remembrance Sunday march past the Cenotaph, where the new King Charles III laid wreaths to honour the war dead. Every living Prime Minister - each of whom had the chance while in office to help the nuclear veterans, and hadn't - stood in a line, as serving troops, Falkland veterans, blind ex-forces, and war heroes progressed past them.

Thanks to the 70th anniversary, nuclear veterans were placed at the head of their column to march. Boniface wrote: 'They began to gather on Horse Guards Parade at 8am, as the autumn fog began to lift. Veterans of the Falklands, the raid on St Nazaire, Afghanistan, Iraq, and the Korean wars met with friends and comrades, berets crisply placed, blazers brushed, medals clanking... There was just one group of men I could see who had no crosses, clasps, or campaign medals for the unique service they alone gave - the 12 nuclear veterans who formed up at the head of column D.' They marched instead with their Missing Medal badges, in a defiant

DAILY Mirror

We Will Remember Them

Friday, November 11, 2022 £1.10 (88p to subscribers)

£5 FREE BET online with BetUK

Valid on Cheltenham plus much more!

SEE PAGE 16

PLUS racing pullout

THE BOYS OF 22

England World Cup squad named

SEE SPORT

SUPPORT George with mum Dinah

SEE PAGES 14&15

Proud of my Boy George

Mum after star's Hancock fury

EXCLUSIVE: NUKE TEST SCANDAL

HEROES DENIED TRUTH

FIRE FLIGHT Pilot prepares to fly through H-bomb cloud

BY SUSIE BONIFACE

NUCLEAR test veterans' blood records have been kept hidden for 70 years, the Mirror can reveal.

The Cold War heroes want a public inquiry into whether they are victims of a criminal conspiracy by the state. The missing blood results meant veterans, such as Terry Gledhill, right, were denied proper care.

As Britain marks Armistice Day, Labour demands Parliament debate the scandal.

FULL STORY: PAGES 4,5,6&7

Veterans still fighting to get full medical records 70 years on

The *Mirror* front page from Armistice Day 2022, revealing for the first time that nuclear veterans' blood tests had been hidden from them for 70 years. Inside, politicians called it a 'criminal cover-up on an industrial scale'. (Reproduced with the permission of Reach plc)

salute to their comrades, and a middle finger to the powers-that-be. The following week, Long-Bailey and Hayes formed a cross-party tag team at Prime Minister's Questions to demand the medal and blood records. 'Under review,' came the response. Had all this work really been for nothing?

The massed ranks of nuclear veterans and families marching past the Cenotaph on Remembrance Sunday, 2022, wearing the Missing Medal where their honour should be. Left to right, Michael Dillon, Colin Moir, Barry Pearce, Terry Quinlan, Brian Unthank. (Susie Boniface)

Eleven days after the march, the veterans gathered at the National Memorial Arboretum for the commemoration promised by Boris, paid for by Liz, and seemingly forgotten by Rishi. Three PMs in as many months, another refusal and a review. They could have been forgiven for not turning up at all. But that morning, Boniface took a call in her hotel room. It was No10, telling her the PM was on his way. In the next breath, they said he would be announcing a new medal for the nuclear veterans. After jumping around the room a few times, she rang Owen, who was on the motorway, driving up with his sister. Once they'd pulled over, Boniface gave them the news. They whooped and hollered like teenagers. At the arboretum the news of the medal, and who would be coming, had to stay a secret. As the rain fell, 200 veterans and families gathered under a marquee by the nuclear tests memorial paid for by the BNTVA many years earlier, and decorated with flowers. There was a ripple of 'ooh's as the PM arrived, with ministers for veterans and defence in tow. Short speeches that had been thoroughly vetted for any controversy were given by carefully-selected veterans, before the PM stepped up.

'At the height of the Cold War, often unseen and unsung, you were in the forefront of our defence. You travelled thousands of miles to the other side of the world to conduct these tests and in doing so you maintained our scientific and technological advantage in the face of the threat posed by the Soviet Union, preserving peace and ensuring the security of our country. The importance of that contribution cannot be overstated. A protective umbrella of our independent nuclear deterrent continues to keep us safe today,' Sunak said. 'But your service deserves more. So I am very pleased to announce today that His Majesty the King has decided to recognise that service formally by creating a new medal for those who served at the nuclear tests. It is a fitting tribute to the incredible contribution that you have made. It is an enduring symbol of our gratitude, so to all our nuclear test veterans, including those who are no longer with us, and all the families that have supported them, on behalf of a grateful nation, I say thank you.'

The announcement was greeted with a sigh, followed by applause, and then tears. Anna Smith, whose husband Barry was refused a war pension three times, said: 'I will bloody well get one. It's only a medal, but it's recognition that he existed. That he fought after the event to get justice, and he was denied it.' Rosie Clark was in tears thinking of her husband Mick. 'It's recognition for his being there, and for all the years we lost. He didn't see his grandchildren, I lost him 30 years ago... It won't bring him back, but at least it's acknowledgement of what he sacrificed.' Laura Morris was asked to shake the PM's hand in a formal line-up afterwards. She told him: 'This is great, but there's so much more that needs to be done.' Sunak promised her a meeting to discuss the blood revelations, war pensions and more. It took six months and multiple reminders before he finally wrote to her about it - saying he was too busy.

Within a month, the new King's correspondence secretary wrote to veterans saying he was pleased as punch to give the first medal of his reign to them. Campaigners immediately asked officials if the medals could be handed out by the King in a special ceremony. The request was repeated privately to ministers, and in meetings over the next six months between LABRATS and government, who had finally decided it was worth talking to them. In January, the new monarch let it be known he was keen on the idea. A source close to the King told the *Mirror*: 'His Majesty wants some form of official ceremony as soon as possible. He recognises that those who can be given it are not going to be around for long. His staff have pencilled in something for April, depending on government approval.'

It was at exactly this point that everything slowed down. Civil servants announced the medal would take more than a year, then spent five months deciding the criteria. It wasn't helped by divisions in government. Veterans' issues had historically been part of the MoD, but it lost territory when Mercer was allowed to set up his own department in the Cabinet Office. He was now answerable only to the PM, rather than a junior minister who had to kneel to the Defence Secretary. There was a 'bitter turf war' between the two departments, and the medal got bogged down in the middle.

Applications were opened in March, with delivery promised by summer. But it soon turned out that it wasn't going to be available to everyone. Those who helped with underground US tests in Nevada up to 1991, or teams of scientists who developed the bomb, including those who designed the earliest weapon systems at Fort Halstead, were barred, along with the RAF crews who had been ordered to fly through the clouds of other nations, and were no different to other nuclear veterans.

John Cammish was a ground crew chief on 543 Squadron's Victor bombers which took samples and photographed the mushroom clouds of French and Chinese tests in the Pacific in the 1970s. Stationed in Guam and Peru, they were kept at a moment's notice to take off on perilous missions aimed at ensuring the British knew exactly what their enemies and allies were doing. In 2011 John had told the *Mirror* how he and most of his unit got throat and tongue cancers. During the interview, he'd had to speak through a tracheotomy after his own larynx and thyroid were removed to prolong his life. Now his widow Lucy said: 'We felt cheated. We wanted to travel and enjoy our retirement. He had to be fed through a tube and developed Alzheimer's. It makes me cross to think of the time we could have had together. He was the first one to touch the planes, open the hatches, offload the equipment, and he had no protective clothing. He was just as exposed as the men at Christmas Island, yet he's not being treated with the same respect.'

The date of expected delivery of the medal shifted from summer to early autumn, and there was talk of protests at Remembrance again if there was no medal in time. The prospect of a show of disgust by old soldiers seemed to unstick the blockage, and Mercer told Parliament he was 'straining every sinew'. Meanwhile, the King had already presented gongs to troops who took part in his mother's funeral, and the coronation. Officials told LABRATS they were still 'actively exploring' a Royal shindig, but then stopped having meetings with them.

Trying to hurry things along, in April the *Mirror* reported the King might be entitled to wear the nuclear medal in honour of his father's visits to Christmas Island, on government business. But it was July before the medal design was revealed, August before the first one was made and the end of September before any veterans received it - in the post, in what felt like a calculated snub. The first to get it was Antler and REME veteran Bob Last, aged 92. His daughter Sandie Catlin said: 'He knows he has dementia, he never smiles. He has falls and other problems. But when he opened the envelope he was so chuffed, he just smiled and smiled. We haven't seen him smile for so long. It's getting that medal which made him happy for the first time in years.' After finishing National Service Bob married and had two daughters. Sandie had cancer three times, her son has seizures, and her sister was born with a curved spine. And Bob could remember blood tests being taken from him when he was at Maralinga. Sandie said: 'He can remember everything about the tests... we made a data protection request for his records, and all we got back from the AWE were nine lines detailing his enlistment, service, and rank.'

There were, though, enough medals sent out quickly enough that those marching at the Cenotaph that year were able - for the first time in seven decades - to wear a medal for their service, just like everyone else.

The medal campaign had taken five years, and four Prime Ministers. Doug, Sandie and Shirley, who had held court in Parliament at the start, had all died without the honour they deserved, just like thousands of others. And there was no party at the

Right: John Morris and his granddaughter Laura, pictured on Horseguards Parade on Remembrance Sunday in 2023 before John marched past the Cenotaph proudly wearing his medal for the first time.

Below: Veterans line up on the parade ground before marching along Whitehall. Left to right, Eric Barton, Terry Quinlan (both seated), John Morris, Terry Hughes, Colin Moir and Brian Unthank. (Susie Boniface)

palace, even though the King was the son of a test veteran himself. At time of writing, 2,649 veterans from the UK and across the Commonwealth have applied and been approved for the nuclear medal, along with 1,825 next of kin, in 25 different countries. They have been given pride of place on more than 5,000 chests and mantlepieces, often next to a photograph of a handsome young man who never lived to hold it.

With the medal delivered, focus moved to the medical records, and by now it had a name: the Nuked Blood Scandal. The MoD was also in a bit of a pickle.

When the question about blood tests was first asked in 2018, it was answered by then-Veterans Minister Tobias Ellwood, who said the MoD was 'unable to locate any information' about it. Presented with the evidence by Boniface, he now asked his successor the same question, and received a different answer. Murrison told Parliament servicemen 'may have had blood samples taken', but they'd be in individual military files. He added: 'The AWE holds copies of the results of urine radioactivity measurements and blood tests for a small number of individuals where these were included in scientific documentation on the nuclear weapons trials.' Why blood tests were needed, he did not say.[43]

In a subsequent answer he contradicted himself, saying: 'The AWE do not hold the blood test results for nuclear test veterans. AWE holds a small number of references to blood and urine tests...' At the same time, the MoD's navy command wrote to the son of a Grapple sailor, saying: 'Although blood and urine tests were due to be conducted, the AWE do not hold any evidence that such tests ever happened. The AWE has come across a handful of ad hoc references within their archives to blood and urine tests... however they do not hold any medical or service records.'

A few months later Murrison repeated the claim that AWE held only 'references' in a letter to a fellow MP, adding there was a record of a blood test only for 'one identifiable member of the armed forces' - presumably, Terry Gledhill. Yet 10 days later, the same minister wrote to a different MP, saying the AWE had 'blood test data for a small number of individuals'.

It was clear the MoD did not know what to say, so Boniface peppered the AWE with requests for clarification on how many blood and urine tests it held, who they were taken from, when, and what analysis had been done. In October 2022 the AWE told her it had 'copies of the results of blood tests, where these were of particular interest, for a small number of individuals'.[44] A month later, it said this was just 'one blood test, for one member of the service personnel'. It added 20 troops had urine tested at Hurricane, but 'AWE does not hold the results'. Two months after that, it admitted having a report about blood tests from Geoff Dhenin and his aircrew who had flown through the Totem clouds before Penney told them to stand down, and that unit medics were told to keep 'a special register in which will be recorded each successive blood examination'. The MoD later said it had no information about any 'special register', and began answering Boniface's questions by directing her to copies of the Gledhill memo and Maralinga orders on the *Mirror's* website, sending her in circles.

Parliament had now been told the MoD held some, one, and zero blood and urine tests of nuclear veterans. Boniface had been told that the AWE held just two documents about it. Veterans who were making their own FOI requests were told although there were orders to do it, they had never been followed. To one veteran, the AWE said it 'would like to take this opportunity to confirm that AWE does not hold the medical records or the results of blood and urine tests for current or ex-servicemen'. The MoD press office said as background that, if anything was held, none of it was 'individually identifiable'.

All the responses ignored the fact AWE, and therefore the MoD, held at least seven blood tests from one individually-identifiable serviceman: Terry Gledhill. In Parliament, as MP after MP asked how a veteran constituent could access his own results, they were told to make data protection requests to AWE for a search of its archives, even though they were apparently empty and there was no personal data. The MoD could not decide which way was up, and then two 'cloud flyers' came forward to say their medical records had been tampered with.

Squadron Leader John Robinson piloted Canberras through the clouds of Buffalo, Antler, and Grapple. When he asked for his RAF medical records, they were provided with 11 years simply missing - from July 1953 to December 1964, covering his time at the tests and immediately afterwards. AWE sent him a dose record equivalent to having 576 chest x-rays in a single year, and told him his thyroid had been tested for iodine, which he was unaware of. John said: 'I don't really remember having any blood taken, but if the orders were given then we would all have done it without question. It is odd there are none there.' Airframe fitter John Folkes, then just 19, was ordered into a Canberra moments before it flew through the Marcoo cloud. Strapped into the 'rumble seat' behind the pilot, he was

Terry Gledhill, on the right, sees off some sailors and civilian visitors from the Christmas Island airfield in late 1958. (Courtesy of the Gledhill family)

told to use an instrument panel to remotely open valves in the wing tips to collect samples. Yet when he asked for his medical records, a 14-month block covering his time at Maralinga was missing. 'I had blood tests taken before, during and after every bomb. Yet only one is in my records - the one I had before I was exposed... the fact they are missing is, I think, highly suspicious,' he said. 'They've since told me I had a zero radiation dose, but seeing as I flew through four nuclear weapons, needed decontamination every time, and at one point had to go into the crater, I'm not sure how they worked that out.' He went on to develop skin lesions on his hands and prostate cancer, found in the latest study to be 34% more likely to kill air crews.

This was exactly what McGinley and Egan had argued to the ECHR a quarter of a century earlier; that medical records had been filleted, and information withheld. Then the Strasbourg judges had decided paper records were bound to get lost, and the case was speculative. Yet the same story was told by others: Dave Whyte, who'd seen his mate buried at sea after Grapple Y, had spent years battling the MoD over the correct readings and whereabouts of his dose badges. He witnessed five bombs, drove into the forward area to collect equipment, seeing his truck decontaminated on the way out while he was not. He lost all his teeth three years later, was found to be sterile in his 30s, and had stomach problems. The MoD refused him a war pension on the grounds he could not have had more than the standard background dose, while Dave's own maths made him suspect it was 30 times greater. Before he was banned from making any more FOI requests by the MoD because of his 'unreasonable persistence', he got copies of his medical records. They showed just one blood test, although he remembers having them frequently, and the signature on his medical discharge was not his.

Arthur Dixon, a fellow sapper, had the job of switching on and sandbagging generators at Christmas Island. He recalled having decontamination showers, and several blood tests, but again his medical records hold just one. The tests for both Dave and Arthur were taken by the same medic, one day apart, using the same forms with spaces for multiple tests, between Grapple Y and Grapple Z. Both show anaemia, and suggest routine testing for the Royal Engineers at Christmas Island, but without the rest of the blood tests, they are useless. Dave said: 'None of us had special clothing. No respirator, no suit. Sometimes you'd put a shirt on, but usually not. Sometimes they'd decontaminate the truck you were driving. We got film badges to measure our dose, and we handed them back every trip, but I've asked for mine repeatedly and they reckon they're lost.'

Evidence the tests *had* been done was mounting. On top of the Gledhill memo, the *Mirror* had found 200 pages of repeated orders for blood counts between 1952 until 1959 with 'safety limits' below which they'd be sent home, instructions for them to be added to medical records in 1955, the form they were to be recorded on in 1956, and a 1957 document claiming 'blood examination' data was being compiled at AWRE with a 'master record held' in the MoD. In 1958, one of Wilf Oulton's officers wrote a memo stating 'certain personnel... are liable to be exposed' at Grapple and asked for blood tests of all air crew, ground crew, and decontamination parties, as well as 25% of other trades. A month later, the AWRE was asked to test servicemen's urine for strontium-90. Officers asked top brass and ministers for guidance, and the paper had even found the results of blood screening conducted before a RAF technician was sent to the Pacific, which showed abnormality and meant he was kept at home instead.

If the MoD was right, and the Gledhill blood tests were the only ones that existed, then repeated orders for blood testing of men in the army, navy and air force were simply ignored. But was there really a spontaneous outbreak of mass insubordination, across all three armed forces, on the most important military and scientific mission of the 20th century? And no-one got court-martialled for it?

Owen summed it up: 'It is inconceivable that with all these orders, and thousands of men involved over more than a decade, there isn't a warehouse somewhere filled with the results. We understand they were held on microfiche at the AWE in Aldermaston, and may have been recently reclassified or moved. We are certain these records exist and are being withheld, and the only possible reason to do that is to limit compensation claims to those injured by the radiation the government has always denied they were exposed to.'

South Shields MP Emma Lewell-Buck, who sat on the Defence Select Committee, said: 'There is enough evidence to show blood tests were ordered, arranged, and taken, from large numbers of people. The results were stored and analysed. The veterans have always had a right to that information, and failing to provide it can cost lives. We must find out when and why they were removed from the medical records.'

On the 66th anniversary of Grapple X, the *Mirror* published a letter written by 22-year-old naval rating Neil Haberman home to his sweetheart Rosemary Jessop. After being ordered onto deck to witness Short Granite, he told her: 'It was a fascinating, yet terrible sight to witness. Fascinating to watch the great, at first, blinding fireball losing its ferocity while developing an ever-growing density of blood-red, as if of that of its anticipated countless victims... Yet terrible when the mind adjusts itself to the enormity of the vista before it, and runs havoc in the trap of visions of such destructive powers unleashed and unchecked.'

For that and two more bombs Neil had no protective clothing, and died of bowel and liver cancer aged 67. His and Rosemary's son was born with brain damage and suffered childhood seizures before taking his own life at 41. Daughter Loraine was born with a blood disorder, had two difficult births, one son with a carcinoid tumour removed from his lung aged 16 and a second with rare genetic condition de Grouchy syndrome, who was never expected to walk, talk, or live past the age of 10. By the age of 34 he was living alone in supported accommodation, but had severe learning difficulties. 'I was born 15 years after that bomb exploded, yet it has blighted mine and my children's entire lives,' said Loraine. 'There's not a single person in my family who hasn't got something wrong. This has haunted us. I know it can be just plain bad luck, but not knowing whether it was the radiation has been an extra burden... the not knowing, the lack of honesty from the MoD, is unforgivable.'

In the Lords, Watson asked the government to help elderly veterans navigate the MoD's labyrinthine archives. But the government response was blunt: 'No

Naval rating Neil Haberman, pictured right in a photo he sent home to sweetheart Rosemary to remember him by, took the photo on the left, of an unnamed friend with the Grapple X cloud blossoming above his head. (Courtesy of Loraine Hawker)

information is withheld... in the case of AWE, they may need to make a Freedom of Information request.'

This was becoming Kafka-esque. Blood tests had been ordered, taken, analysed and stored but the only means of finding them were for sick men in their 80s to find a way through legislative minutiae which, if successful, produced medical records that were missing the relevant information. It was the same sort of circular demand made by the MoD when it asked veterans to produce evidence of their injury - evidence only it had ever held. And the MoD was frank in refusing, point blank, to produce what it had. FOI requests make information public, and to do so with a medical record would breach patient confidentiality, said the MoD. In those cases where a veteran or next of kin got confirmation there were blood tests on file, they were not allowed to see them. Gledhill's daughter Jane was told there were further blood tests, and after multiple appeals the decision on whether to reveal them had been made by an unnamed minister. That politician, whoever it was, had blocked publication, and the MoD refused even to share the advice he had been given by officials which led to it.

As far as the *Mirror* was concerned, these were crimes committed by the state against its own servicemen. It had been made a criminal offence to withhold, falsify, or destroy medical records in 2014, and it was always a civil wrong which could see the person or organisation responsible sued. Any medic involved ran the risk of being struck off by their professional bodies and barred from practicing medicine for life. With many military doctors and nurses regularly seconded to, or trained by, the civilian National Health Service, it was simply impossible that they would be allowed to falsify a medical record in one job, but face legal penalties for doing so in the other.

Politicians, including an ex-PM, an ex-Health Secretary and the Shadow Attorney General were in agreement. If the MoD was holding out, it was clear the veterans would need either a police investigation, or a lawyer.

Boniface contacted law firms with a reputation in the field, but made little headway until Watson put her in touch with Jason McCue, a human rights specialist who, with partner Matthew Jury, was already taking on Russian oligarchs in the wake of the war in Ukraine, and had a track record of victim-led group actions. The peer's intro got her a Zoom call with them, in which Boniface explained in detail the crazy situation veterans found themselves in, told to ask for blood tests that could not be found. McCue asked Boniface what the veterans needed, and she replied that while it was different for each of them, they had a black cloud hanging over them, and it needed to be lifted. McCue was struck by inspiration, and laughing said: 'Well you know what we do - oh, this is brilliant, this will work - we sue them for the medical records they say they haven't got. The only way to fight the case is to produce what they're hiding, or else say what happened to it. If the records have been lost, they have to pay up. If they've been destroyed, they have to pay. And if they produce the records, then if they show damage there's a payout and if they show no damage there's still a payout, because it was illegal to withhold them. It's Hobson's choice, we get them coming and going.' Even without blood tests the veterans could still claim damages for post-traumatic stress, which, unlike radiation, had been legally proven to infect whole families for generations.

There might finally be compensation, but it had to be paid for. While a funder would probably come on board for the final and most expensive stages, the case would most likely need at least one High Court hearing so a judge could order the MoD to disclose its records. That required a barrister, legal prep, and would cost somewhere between £50,000 and £100,000 which the veterans would have to raise themselves. When the campaign steering group met with the lawyers to hear for themselves, the hope of legal victory was tempered by the problem of raising such a huge sum. History showed the MoD's tactic was always to play the ball long, dragging out legal action past the point anyone could afford to keep funding it, and after the litigants' time on Earth had run out.

If there was to be a formal medal ceremony with the King, it would be for many veterans their last chance to be properly honoured. If they sued the government, it would never happen. Campaigners decided to hold fire, again, until it was confirmed there would be no party at the palace. Then on September 21, 2023, they detonated their case with yet another *Mirror* front page: 'We just want the bloody truth,' said John Morris. Inside, across two pages, there was fresh evidence: the AWE had admitted holding at least 150 documents about blood and urine testing during the weapons trials. Only their titles were known, including 'blood test data', 'special medical examinations', 'decontamination and blood counts', 'blood changes at low dose levels' and even 'report on medical examinations of natives'. They spanned 1952 to 2002, and showed 'how servicemen at the nuclear tests were used as guinea pigs and routinely given blood tests and medical examinations to measure levels of contamination'. McCue said the veterans 'have been gaslighted by the British state... they were treated like spent ammunition, discarded and disdained by those who had a legal and moral duty of care'. He added that, if civilian doctors had access to the records, it might have altered diagnoses and treatment, and 'there might be many more of them alive today'.

For the first time the campaign went international, as Boniface teamed up with *Adelaide Advertiser* reporter Andrew Hough to simultaneously expose the blood scandal to Australian survivors. The same day, the BBC interviewed Eric Barton about his missing medical records, in a triple whammy of media coverage.

The AWE insisted that things were not how they looked. 'We would like to highlight that some of the document titles may be ambiguous as to their content,' said a spokesman. The newspaper highlighted the crowdfunder and asked readers to donate, with the announcement that its parent company, Reach plc, had put in the first £20,000. Brian Unthank, John Folkes, Steve Purse, John Morris, and Laura Jackson became the first to sign up, and the next day, at the veterans' annual reunion in Weston-super-Mare, more than 100 others joined too. Barbara Butler, whose husband Tony had died riddled with tumours, said: 'Tony's watching. I know he's happy about today. It's a day we thought might never come.' She dried her eyes,

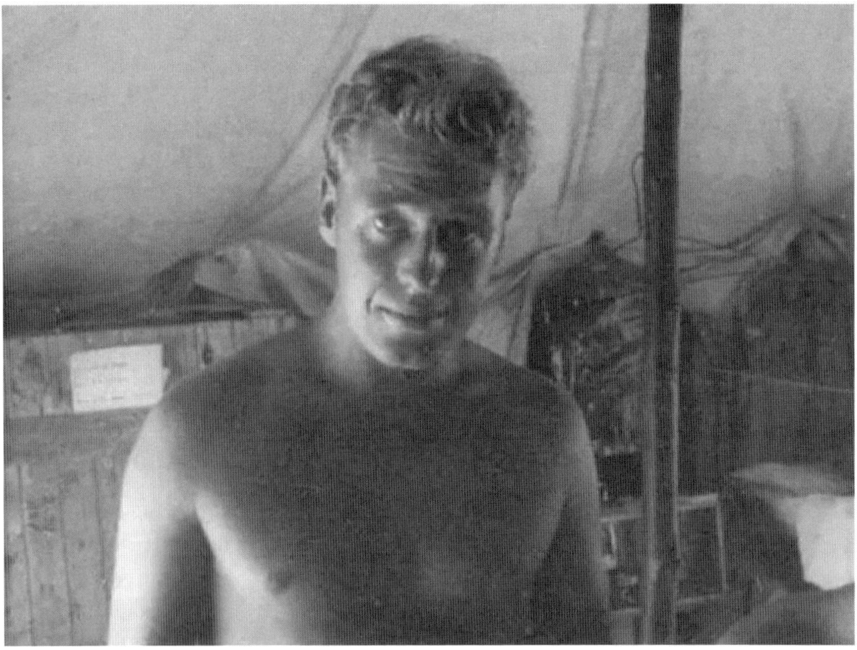

RAF corporal Tony Butler, pictured during his time at Christmas Island. He suffered skin cancers and tooth problems for decades before an MRI discovered he was 'riddled' with tumours. He died aged 72, in 2009. (Courtesy of the Butler family)

looked to the sky, and said 'This is for you, Tony.' Her daughter added: 'Everyone's got a smile on their face now. We've got hope.'

Government remained intransigent. It refused the request of peers and MPs to provide copies of the 150 files. Despite being known as a veterans' champion, Johnny Mercer also declined to help. Owen presented him with 200 pages of orders and discussions about blood testing, as proof something was amiss. 'I told him that not having this information meant veterans weren't getting the right medical treatment today, on his watch. He skimmed it and said "If you've got the evidence, you need to start a legal case against the MoD." On my way out of the meeting one of his officials said "Please don't make the headline Johnny Mercer Tells Veterans to Sue the MoD".' The government confirmed the meeting, and that those words were used, so that was exactly the headline the *Mirror* ran. A government source told the paper that Mercer had received reassurances from the MoD, and when he was later accused in Parliament of a 'dereliction of duty', he was furious. 'There is no cover-up,' he said. 'Some records were taken, some were not. There is no cover-up to discriminate against that cohort. It simply does not exist. What would be the reason to cover up rather than look after these people?'[45] He accused his opponents of politicising the test veterans, but did not discuss whether the wish to avoid compensation, and humiliating apologies to Commonwealth allies, might explain it.

Britain was entering the final year of a government mired in failure and scandal. Labour was comfortably ahead in the polls and widely seen as likely to take power. Its press operation became more cautious, with MPs forbidden from making promises that might turn out to be irreversibly expensive. So when deputy leader Angela Rayner was invited to speak publicly about the blood records with John Morris at the 2023 party conference, no-one expected more than warm words.

John spoke about his decades of trauma, about his anaemia, the blood tests, and the fact a war pension had been refused for his work in the Christmas Island laundry, the same one whose water tanks and washing machines were later found to be so contaminated they were dumped at sea. 'After my son died, every morning after we had other children I had to get up first and make sure that child was breathing, before my wife would get out of bed. You carry that,' he told the party conference in Liverpool, to a hushed auditorium and barely any dry eyes. 'How can my doctor, any hospital, treat me when they don't know what they're treating me for? Both of my daughters have had miscarriages. One of my daughters has been told it would be better if she did not have children... the cover-up has got to stop.' He made four requests: a public inquiry, the missing records, a Hillsborough Law to improve accountability, and compensation for those affected.

Rayner had arrived with a sheaf of notes, and did not refer to these official lines-to-take once as she leaned across and spoke to John directly. 'It's really difficult to hear what you've been through, what your family's been through, and still not to be given the respect of the truth, I think it's disgusting... after hearing what you've said today John I'm going to push hard to meet your four asks and to do everything I can to make sure that, in your lifetime, I'm able to give you that.'

More than four decades after the scandal broke, with the most evidence ever collected of an official cover-up, and the most hopeful legal case they'd ever had, it was the most that any politician had ever promised the veterans. It was, though, still just a promise. Each of the follow-up stories was accompanied by more calls to donate to the cause, and the crowdfunder grew from public donations, with veterans themselves putting in what they could afford. But at £35,000, it still wasn't enough.

Labour donor Dale Vince, a former hippy who had founded green energy company Ecotricity, had the bad luck to walk past Boniface, Owen and Purse outside the event. They grabbed him and asked for his support, and a few weeks later he came good with a £15,000 donation, to get them as far as the court steps.

As hundreds of medals began to be delivered, and veterans and families resorted to organising their own presentation ceremonies with local Lord Lieutenants, mayors and MPs, the PM refused to comment on the case. Then the *Mirror* revealed the AWE had admitted it might hold almost 5,000 personal files with blood tests in them. Boniface explained: 'We asked the AWE... to define exactly how many veterans' blood tests it held. At first we were told there were a "small number", then "one blood test", and that there were no blood records in its archives. After a complaint to the Information Commissioner, the AWE admitted it holds 4,711 files "which may contain references to blood or urine testing". It said each file would need to be individually examined to determine its contents, and at an estimated 15 minutes to check each record, it would exceed the cost limits set under FOI laws.'

That November, 45 veterans marched past the Cenotaph once more, this time wearing their new medal. For John Williams, 81, a survivor of Operation Dominic, it was extra special. He had always watched from home so it was his first visit in person, but with bad knees and a lung disorder he simply could not keep up with the rest of the contingent. He fell behind, and those marching next in the procession slowed down rather than risk overtaking him. A 200-yard gap opened up between John and the test veterans in front. With his son Keith holding one arm and Boniface the other, he was applauded by onlookers and told: 'Keep going!' As he slowly passed Princess Anne, alone, she gave him a personal salute. 'I've done it!' he said afterwards. 'I always wanted to be here, and now I have. With my medal, too. It's been just wonderful, and I'm so glad I did it without a wheelchair.'

Also there was Terry Washington, 89, who was aboard HMS *Warrior* for Grapple. He wore his new medal on his left breast, and on the right, the Missing Medal. 'It's just as important as the other one. It's like our National Service medal - 70 years we've waited for this,' he said. Michael Dillon, 84, said: 'I'm overjoyed at getting the medal, but I just wonder why it was so long that we had to wait for it. That is now very nearly 65 years. It was 1960 when I went to Christmas Island.' Asked what difference the medal made to him, he replied, 'Every difference, my love', and welled up, unable to speak. Bryan Jarvis, 85, was a 'nut strangler' - military slang for those in the REME - and glad to wear his medal. He added: 'My medical notes from Christmas Island are almost totally missing - the only scrap of paper left says I had sunburn. Well, now I've had eight skin cancers. We were microwaved out there and they won't admit it.'

That morning John Morris, having breakfast in his hotel, was approached by someone who'd noticed his medal was the new one for test veterans. 'The medal makes me recognisable whereas last year I was a nobody,' he said. 'I wear this medal with a great deal of pride, and I wear it for the 20,000 plus who are now dead and never saw it.'

The medals were still arriving, and veterans arranged their own fanfare. Colin Moir, whose idea it had been, was presented with his from the local Lord Lieutenant. Geoff Colgate, 88, who had winched the Grapple Z balloons into the air in 1958, had a visit at his care home from the air commodore of his local RAF base. And widow Margaret McCann, who had lost her husband Frank half a century earlier, was pictured in the *Mirror* on her bed in a hospice, with the medal pinned to her nightie and a smile on her face. Frank, known as Ginger, was a sapper who died at 34 after suffering two strokes, nephritis, and a massive heart attack a decade after his return from Christmas Island. His wife had two miscarriages. Daughter Fran said: 'We wanted the medal because for us it's recognition he was there, and those dreadful things happened. She's at end of life now, but she was really thrilled when his medal arrived.'

Clive Osborne, 86, got his from the postman. Daughter Karen had told Adrian Davies they were waiting for a special delivery, and when it arrived he agreed to step inside and pin it on the former RAF clerk. Gordon Lowe and Dennis Leafe had their medals placed with care next to their photographs, while Arthur Coe put his on the windowsill next to a picture of his mates from the RAF, who had all passed away.

The medal was won. The crowdfunder was growing, and the case was being built. The AWE admitted in another FOI that the 150 blood and urine documents were held on a secure database called Merlin, and locked from public view under a Lord Chancellor's Instrument, an order from the government's top lawyer. Each LCI has a schedule of the documents to be covered, and a brief explanation of the reasons. The LCIs covering these documents, though, were highly unusual. The schedule made no mention of AWE files, and each LCI had been signed several years earlier than the date on which the records were locked. It seemed that someone had hidden them by using the LCI retrospectively, adding documents to it without any public note being made.

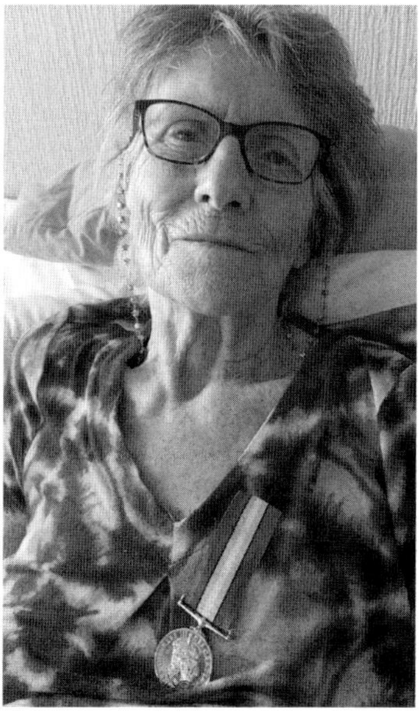

Margaret McCann, pictured relieved but happy in her hospice, finally wearing her late husband Frank's medal. (Courtesy of Fran McCann)

Now the screws were tightened on the government, with a special Parliamentary debate secured by Long-Bailey. Held in a committee room rather than the main chamber, it did not draw much attention. It did, though, demand an appearance from a minister to respond to the motion 'that this House has considered nuclear test veterans and medical records'. Andrew Murrison, who had so often repeated the contradictory government lines, took part in what turned out to be a more detailed, calmer and cooperative discussion than the test veterans were used to seeing.

Long-Bailey said veterans 'and sometimes their wives, widows and descendants have reported making repeated requests to gain access to the results of their blood or urine testing samples which they recall being taken during the nuclear testing programme. Sadly, many confirm that their service medical records do not include the test results, and they just do not understand why... were blood and urine samples taken from nuclear testing veterans, and was a record kept of those samples?' She listed the evidence gathered, including documentary proof the MoD had a 'director of hygiene and research' who kept a 'master record' of the results, the orders given to unit medics and pathologists, and that thousands of documents that could contain clues were pulled from the archives in 2018. She quoted the titles of some of the 150 Merlin files, as well as the MoD's merry-go-round of excuses, some of which had been signed by Murrison himself. She asked him to personally review the security classification of any files, re-release the records withdrawn from public view, and give an explanation about what had happened to the blood tests that were undoubtedly taken.

She was backed by MPs of all parties. Democratic Unionist MP Jim Shannon said: 'We are all asking: is there a cover-up?' Emma Lewell-Buck pointed out the NRPB report 'had been tampered with' and spoke of a constituent, Jack Taylor, who served at Antler and had shown her 'mountains of dismissive letters from various Secretaries of State and ministers'. She was followed by Sir John Hayes, who said: 'The publication of records from 70 years ago is being withheld on what are, in my judgment, highly questionable grounds. I say that and I use the term "cover-up", which is not one that I would deploy lightly, because the MoD initially denied that there were records at all. For years, we were told that these tests were not done... It has been only through Freedom of Information requests that we have discovered that blood records do indeed exist held at the AWE.' He said some documents showed the blood tests were missing from medical files 'as early as 1959' and added: 'If there are tests, how many are there? Where are they held? Was the analysis done, and if so, does it still exist in a way that can be scrutinised? Were the records generic or particular – were they maintained and named to individuals? Those are all matters that I know the minister will want to make this chamber aware of without delay.'

Walthamstow MP Stella Creasy said it was 'a mark of shame' that veterans were having to take legal action to get the truth. 'We shouldn't be waiting for people to make requests for their own medical information. There should be no question about whether data can be released. We should be humbled and horrified enough to get that information to them, and proactively investigate the healthcare concerns they and their families may have,' she said. Shadow Armed Forces Minister Luke Pollard said, 'It does not sit well with me' that veterans had to fund a legal case to get the truth, and asked Murrison to give them peace of mind.

All eyes turned to Murrison, a former navy medic who, despite a reputation as a diligent politician who played a straight bat, had relied on officials to compose letters and answers on his behalf which had not been up to snuff. Political opponents in the Labour party, right wing allies and even some from his own backbenches were united in saying something was off, and he had been put on the spot. What happened next must have been considered in advance.

He said that some of the claims risked 'unduly alarming' veterans, and repeated the line that the AWE held no medical records. He referred to the possibility of cover-up, not as impossible, but 'curious', and then turned to the list of 150 blood and urine documents. 'One referred to an anonymous blood test, another contained four anonymous urine tests and the last identified one individual's blood tests,' he said, producing a total of at least six individual diagnostic tests which the AWE never admitted to. 'I have looked at the subject headings and asked officials to look again at the 150 files with a view to placing those not already available to the public in the public domain. I have also asked to see them myself,' he said.

He had been told by officials that the one identifiable test the *Mirror* had found was the only one that existed. 'The AWE has not been able to tell me specifically why there should be that one case, out of all the data they hold, that is personally identifiable... I don't know why that is, it could have been a mistake, I just simply don't know. That's why I've asked to see those 150 files myself,' he said. 'Some are tantalising and refer to test results. I would like to see what those documents look like. I have not seen them so far, and I certainly intend to examine them myself... Wherever I can possibly do so, I will ensure that that material is placed in the public domain... if it is simply sheets and sheets of dosimetry and urine and blood test results, I cannot see why that should not be available. I will certainly make it my business to examine that in the days ahead.'

In political terms, the minister had offered to dip his hands in the blood he hoped wasn't there - to take personal responsibility for whatever was in the files, and risk the consequences if he got it wrong. If the AWE documents showed a cover-up, he would be sucked into it if he did not publish them, and if they did not, he would have to publish them anyway. With a legal case pending, an Opposition likely to take power and talking about a public inquiry, there would be no plausible denial if he had seen the records.

Within days he had sent to AWE for the files to be delivered to his office. Meanwhile campaigners were uncovering more. Culture Minister John Whittingdale, whose responsibility it was to keep an eye on public record-keeping, told Parliament that: 'The grounds for retention of those AWE files... are national security, security against possible terrorist activity, international relations... the risk of proliferation of nuclear weapons or to allow a more detailed review.' Sir John said, 'It beggars belief that a medical diagnostic test... is in any way a state secret' while Steve Purse added: 'It's frankly ludicrous to think that giving me that information would endanger world peace. It's far more likely the information they have would lead to compensation, and that's what really terrifies them.'

Three months later, Murrison had to tell Parliament he had yet to see the files, but despite that, 'I think very few of them are actually going to give you any information that's going to take us any further forward'. MPs laughed in his face, while Owen asked the *Mirror*: 'If there's nothing in them, why are they a state secret?' A few days later it became clear why the review was stalled. 'The AWE has been accused of trying to stop a defence minister seeing 150 documents about blood tests taken from British troops during radiation experiments,' revealed the *Mirror*. It quoted a government source who said: 'Andrew has taken a lot of flak from the nuke vets, but he's a decent guy and when he realised these records existed he knew many probably could and should be published. He asked the AWE for all 150 files, plus background papers, to review them personally. But they took ages and all they sent him was a few bits of paper they'd cherry-picked out of that pile he asked for. He kicked it back to them, and is determined to do what he promised.'

It was extraordinary that an agency of the MoD was refusing to comply with a request by an MoD minister, and that it was being leaked to the media. Again, publicity seemed to clear the blockage, and the AWE conceded he could visit, and look at the records in person. Meanwhile the MoD lost on another front, as Gledhill's daughter Jane won a landmark ruling over access to her father's medical records. Her appeal had slowly wended its way to a judge, and a barrister called Joseph Barrett had offered to act for her pro bono.

The one-day hearing had to decide on a loophole in the law. The Access to Health Records Act guaranteed a person, or their executor, a copy of medical records from 1991 onwards. Before that, there was no such right. Therefore veterans from as recently as the first Gulf War were unable to obtain proof of any injury, or pursue any claim. The only possible legal route was an FOI request, but the snag was it meant publishing it to the world. This would be a breach of patient confidentiality, argued the MoD, and must be refused.

Barrett said the effect was to deny records to those who, like Jane, would otherwise have a perfect legal and moral right to them. He pointed out that, as she was executor of her father's estate, and had waived any right to sue over confidentiality, they could now be published. The MoD's main witness was Brigadier Duncan Wilson, medical director to the Surgeon General of the armed forces. He provided a witness statement saying 'it was not clear' what Gledhill's medical records would add to the issue of whether troops were irradiated, and that they should never have been provided to Jane in the first place. Under cross-examination, he admitted the statement had been prepared by officials and he had merely signed it, but that he was no radiation expert.

The judge ruled in Jane's favour. Asked for comment, the MoD gave the *Mirror* a 402-word statement about the medal, and claimed - in the face of a year-long legal battle - that 'it remains the case' any veteran could simply request their medical records. 'I'm over the moon, and I hope dad would be proud of me. As an officer he always wanted to take care of his men, and with this we've set a precedent which could mean other families can likewise see the blood tests of their fathers too,' said Jane. When the papers came through, they were intriguing - showing that Gledhill's doctors were checking him for signs of radiation a decade after Christmas Island, and that his blood never got back to normal.

One thing the medal had changed was awareness. People were once again interested in hearing from veterans, and astonished to find that they had waited so long, been treated so shabbily, and were still finding fresh evidence of a cover-up. John Morris and his granddaughter, Unthank, Owen and others took turns to sit on the TV sofas with each new development, keeping the story simmering in the minds of the public. The *Mirror* could no longer command an audience of millions, and it was vital to get TV cameras to look again at what they had ignored for so long.

It drew interest from documentary makers and as the nuked blood scandal heated up the attention turned into a clamour, with more than one firm demanding exclusive rights to the veterans and their legal case. After years of Boniface getting knocked back when she approached commissioners, it was a welcome change, if annoyingly late in the day.

On March 19, 2024, cameras were rolling for what everyone hoped was the final battle: formally serving legal papers on the government. A day after the *Mirror* reported that Sunak had blanked their requests for a meeting 12 times since becoming Prime Minister, Terry Quinlan and Brian Unthank marched the length of Downing Street to knock on his door. They were backed by two lawyers, Alan Owen, Steve Purse, his wife Georgina and toddler son Sascha.

To ram their point home, Brian and Terry were dressed in the same protective clothing they had on Christmas Island: shorts and boots. As the *Mirror* pointed out: 'These two men have clocked up 92 skin cancers, two tumours, and 13 miscarriages between them since serving at Christmas Island.' Terry had kept his old floppy bush hat, and they both wore t-shirts saying: 'Give us the bloody truth.' They saluted smartly outside the famous polished black door, while half a dozen camera crews snapped away and Owen handed over a petition, on top of which nestled a copy of the legal letter. Afterwards, they did interviews with Press Association, the BBC, and Japanese television before heading on to their old foe: the MoD.

They marched up to the neoclassical building overlooking the Thames which had been the focus of their rage for decades. Built from Portland stone in the 1950s, the Air Ministry moved in around the time of Grapple Z, and today it houses the headquarters for the entire defence department. It stands on the remains of the old

A suntanned Terry Quinlan and pal lounge on the front fender of the Army truck they drove all over Christmas Island, without protective clothing, during the Grapple tests in 1958. (Courtesy of Terry Quinlan)

Terry Quinlan and Brian Unthank salute on the steps of Downing Street, as Alan Owen prepares to hand in a petition and legal letter. Veteran's son Steve Purse, his wife Gina and son Sascha, were also there to call for the 'Bloody Truth'. (Susie Boniface)

Palace of Whitehall, home to monarchs and government since Henry VIII, and somewhere beneath it lie the remains of that mad monarch's wine cellar, still used for entertaining. The MoD has 10 storeys above ground, and another three below. Its basement was built to provide a command centre in the event of nuclear attack, and there is almost certainly a brigadier in it.

The veterans and families formed up on the steps, holding their legal papers aloft as the media gathered around them. Within a few minutes, just as on their last visit, armed police appeared, insisting the veterans complete their business. Owen led his men into the lobby to deliver the letter. They soon trooped back out, as the receptionist had refused to accept it: it didn't matter. The point was to pull a stunt.

Unthank was not optimistic. 'For all the hard work that's been done, I fear this was a waste of time. The MoD will try to sweep it under the carpet again,' he said. Quinlan was more upbeat: 'They cannot ignore us any longer. Whatever happened to my blood and urine tests, they have to compensate and apologise for not giving me them, or losing them. I hope this is the last time we have to demand what we should have had long ago.' The legal papers came with an extraordinary offer to avoid court, and cut the costs as well as the humiliation: a special one-year tribunal, with all the powers of a public inquiry, and the ability to investigate, adjudicate, compensate and commemorate the testing programme. If the MoD ignored the offer, it could multiply future damages.

Just 24 hours later, the government announced that Andrew Murrison had ordered the declassification of 4,000 pages of the Merlin files. The *Mirror* declared that the minister must have found them to be 'politically radioactive' and ordered the release to protect himself from contamination. Just 10 weeks earlier, they were critical state secrets, locked away indefinitely. They were about 'blood counts' that, according to the MoD never happened, and 'blood count data' that the AWE said was never analysed. Some of them, including the Gledhill memo, had been hurriedly pulled from public view when the *Mirror* launched its medal campaign. They had been stored with technical data about nuclear weapons, on a secure database to which only a handful had access. Now they were to be published, 'in due course'.

FOI requests established that Merlin had been created as a secure, standalone database in 2014. At least some of the documents added to it were locked before this date, and they included some of the ES files. To lock the records, more than one Lord Chancellor's Instrument had been invoked retrospectively, without public trace. The purpose, according to AWE, was to hold documents relevant to nuclear veteran litigation, which had ended in the Supreme Court in early 2013. Why information relevant to lawsuits was a state secret, whether its records about blood and urine testing were disclosed to lawyers or how it was that Terry Gledhill's blood tests were considered 'technical data' about weaponry, remain unanswered.

And there was a possibility they always will be. AWE admitted that, while there was an audit trail keeping a careful note of who accessed Merlin and why, a software update meant it had no information about who had been in the secrets between 2014 and 2022. After that, it confirmed Merlin had been accessed 283 times, and the only reason given on the form was 'opening database'. Somehow, the British government's cache of documents with legal and proliferative risks, which it took a minister four months of arguing to access, had been looked at by officials once every three days, yet no-one could say why.

Owen was clear that there was only one reason these documents had ever been uncovered: it boiled down to Richard Stott, and the longest campaign in newspaper history. 'There has never been any reasonable explanation why medical diagnostic tests, the only purpose of which is to find out if veterans had been exposed to radiation, were hidden,' he said. 'That we know about them, and they are now being forced open, is purely as a result of pressure in Parliament and the media.'

The documents were finally published online in May 2024,[46] eight months after Boniface had uncovered their existence. Their contents included proof that more than 1,000 blood tests were taken, medical record forms the AWE had denied having, 30 separate instructions for mass blood testing and at least 550 named servicemen listed for the 'special medical examinations'.

Among the pages was a fax from 2001, in which officials in the MoD and AWE shared some of the documents discussing how the blood tests were ordered and conducted, and then composed 'lines to take' for ministers and the media which said they were never carried out.

The *Mirror* called it evidence of a conspiracy, and Andy Burnham agreed, saying: 'The *Mirror* has got the MoD bang to rights: this is the proof that it has been lying to our servicemen and our country for decades... they could more accurately be described as "lies to take". They were given to ministers and the media when the MoD was sat on a mountain of evidence to the contrary.'

The fax, and the rest of the damning evidence, was hidden behind bogus claims of national security for more than 20 years, as nuclear veterans campaigned and successive governments said there was nothing amiss. If it had been uncovered sooner, it might have led to research, compensation and an apology for all those affected. Lives may even have been saved; we will never know.

But it is clear that if Richard Stott had ever had evidence of blood tests, if the Australian Royal Commission ever saw records of 'gross irregularity' in servicemen's blood, if Shirley Denson had been able to prove radiation was in Eric's veins, if the inquests of Billy Morris, Derek Redman, and Stephen Pooley ever had the results of samples taken at Christmas Island, then the campaign might have ended long ago. If the ECHR and Supreme Court had seen the evidence of blood tests which has been uncovered in the past two years, the rulings of each might have been different.

Ninety per cent of the troops at the British nuclear weapons tests have gone. Those who remain, by definition, are the least-affected. But they and their families still search for answers from institutionalised civil servants and incompetent politicians who will not or cannot see that something went wrong, and that it must be corrected.

If they get them, even this late, Steve Purse can still show his dad's blood tests to his doctor. Alan Owen can be told whether his family's problems are simply bad luck. John Morris can stop worrying about his great-grandchildren. Descendants can make informed choices, doctors can make better diagnoses, and all citizens could have more faith in the British state, if the government just told the bloody truth, rather than waited for its wrongs to be exposed.

EPILOGUE

ON A BARE stretch of highway just outside the Outback town of Woomera - which has a population today of just 146 people - is a cemetery. The land for it was acquired in 1953, the same year weapons tests began in the nearby desert. It contains 250 graves, and 71 of them - 28% - are children.

Only one was born before the weapon trials. All have identical plates stamped with their names, ages, and dates of death. There are 22 stillborn babies, 14 who died within hours of birth, and 24 who had died by the age of one. Nine died before they were five, and two died aged seven.

The graves' existence was first revealed in 2003 by investigative reporter Colin James of the *Adelaide Advertiser.* Every child had been born or died in the local military hospital. The official explanation for the deaths is a heatwave in 1959, even though the first grave was dug in December 1953 and the last in September 1968.

The death certificate for each child is sealed in the Australian national archives. James said many former residents could not speak on the record because they had signed the Official Secrets Act, but one woman told him: 'I lost two of my babies… I've never believed all of this poppycock about it being the heat. It might explain

One of the children's graves at Woomera Cemetery. (Daniel Everitt-Lock, True Perspective Films)

the deaths of one particular year, but what about all the other years when babies still died even when there was air conditioning?'

His dossier included reports of doctors' wives from the base going for picnics to watch the mushroom clouds, and later suffering miscarriages; A-bomb documents that were never shown to the Royal Commission; Australian scientists later using the same sites for nerve gas tests; and how medical records from the military hospital had vanished, with Australian servicemen who later fell sick unable to access them.

He wrote about Robert Williamson, who served at Monte Bello, and cut his finger on a later posting to Maralinga. It never healed, was amputated, and then led to cancer which spread up his arm and killed him, aged 44. A son and two grandchildren developed brain tumours, and other descendants suffered miscarriages, cancers and bone problems. His widow Diane said that after her grandson was diagnosed with two different cancers, she realised 'there must be some sort of gene running haywire through the family. Something has happened to us'.

Woomera is just one of the many graveyards where ghosts wait for justice. In the summer of 2024, a snap general election was called. Keir Starmer took power with a landslide of 412 Labour MPs, many of whom - like he and members of his Cabinet - had demanded compensation and recognition for the nuclear veterans while in Opposition. It takes time to turn the ship of state, but there is no sign that the MoD is about to reverse its long-held position of petty-minded cruelty. Within a month of Starmer taking office, the Mirror reported that Sqn Ldr Pete Peters - at 92, the last survivor of Operation Bagpipes in 1954 - had just been refused the medal, on the basis that he did not fly through British mushroom clouds. 'I'm pushing like blazes while I still can. The bloody medal criteria is wrong,' said Sqn Ldr Peters, vowing to fight on for the sake of his comrades and that unheralded crew, lost long ago in the storms of the Pacific.

As this book goes to press, the few remaining veterans still await the fulfillment of those politicians' promises - of widening the medal criteria, delivering the medical records, holding a public inquiry, paying compensation, and most importantly giving them the relief of an apology that is so overdue. Until that happens, the campaign will continue.

Among Churchill's old papers from 1953 is a memo from his Paymaster General Lord Cherwell. He states that in 1944 and 1945, the PM was urged to tell his Cabinet about Britain's nuclear programme, and refused. A few months after D-Day Churchill was told the Americans would have an atomic weapon ready for the summer of 1945, and as his troops fought through Europe and north Africa it was this all-powerful new weapon on which Allied hopes rested.

Cherwell's note makes two astonishing claims. 'As you know it has now been proved that the Japanese asked the Russians to convey an offer accepting the Potsdam term of unconditional surrender to the Allies on August 2. The Russians did not pass on the message. Had they done so, it might well be that the bombs on Hiroshima and Nagasaki would not have been dropped. Whether the fact they were dropped did more than induce the Japanese to make a direct offer of surrender to the Americans is therefore questionable.'

Having blamed the Soviets for sitting on their hands, the PM's friend and former scientific adviser - considered by historians to have been an arrogant man of poor personal judgement - said these horrifying weapons would not alone have been enough to end the war. 'In retrospect, it seems unlikely that the atom bombardment could have forced the Japanese to surrender before the planned invasion [...] had they not - unbeknown to us - been already at the point of collapse,' he said. 'For the Americans could not have dropped much more than two bombs a month for the rest of the year. But of course the Japanese did not know this.'

Two bombs a month, for the rest of 1945, would have meant a total of 10 nuclear blasts. And Cherwell, a physicist who was at the heart of all these international discussions, considered that insufficiently persuasive.

Two bombs were needlessly dropped on Japan to induce a surrender that had already been offered. Since then, total of 2,476 nuclear weapons have been tested by at least eight countries, to create a weapon they all declare they do not wish to use. Tons of plutonium and uranium have been vaporised in the atmosphere, with little to no official analysis of the impact it had, if any, on miscarriages, cancers, or blood disorders of the human race at large. It is this lack of knowledge which causes most fear.

As far as those who have witnessed nuclear weapons are concerned, there is nothing worse. But to those that create and wield them, the threat may never be enough.

Endnotes

Introduction

1. https://time.com/3494421/hiroshima-and-nagasaki-photos-from-the-ruins/

Radiation

1. https://www.osti.gov/opennet/manhattan-project-history/Resources/einstein_letter_photograph.htm#1
2. Test of Greatness:Britain's Struggle for the Atom Bomb, Brian Cathcart (1995)
3. Test of Greatness:Britain's Struggle for the Atom Bomb, Brian Cathcart (1995)
4. The Venona Agency, Robert L. Benson, No date
5. British Nuclear Tests in Australia - Royal Commission, 20 November 1985 - Volume 1, 2.1.11
6. The Seventh Decade: The New Shape of Nuclear Danger, Jonathan Schell, No date
7. British Nuclear Tests in Australia - Royal Commission, 20 November 1985 - Volume 2, 12.3.14
8. Mortality and cancer incidence 1952–2017 in the United Kingdom's atmospheric nuclear weapon tests and experimental programmes, Gillies and Haylock, Journal of Radiological Protection, 2022
9. Testimony to the Australian Royal Commission, 1985, 5134 - 5144
10. *The West Australian*, p1, Oct 4 1952, National Library of Australia
11. *Daily Mirror*, p1, Oct 4 1952, British Library
12. Hansard, 23 October 1952 vol 505 cc1268-71
13. *Canberra Times*, p1, 10 Nov 1952
14. ES20/16, National Archives, early exchange of letters between Lord Butterfield, William Penney and the Medical Research Council 1952
15. PREM 11/563, 1952-1953 Tube Alloys, National Archives
16. COS (53) 239, Chiefs of Staff Committee, Atomic Weapon Trials, report by the Defence Research Policy Committee

17. British Nuclear Tests in Australia - Royal Commission, 20 November 1985 - Volume 1, 6.1.15
18. *South Australia News*, p1, 15 Oct 1953
19. DEFE 16/2, Canberra flight report, October 1953 (Operation Hot Box)
20. Air Marshal Sir Geoffrey Dhenin's obituary, *Daily Telegraph*, 11 May 2011
21. British Nuclear Tests in Australia - Royal Commission, 20 November 1985 - Volume 1, 6.4.19
22. The Black Mist and its Aftermath - Oral Histories by Lalli Lennon (2010)
23. British Nuclear Tests in Australia - Royal Commission, 20 November 1985 - Volume 1
24. Question Time, Hansard, House of Representatives, 21 Oct 1953
25. https://www.nma.gov.au/defining-moments/resources/maralinga
26. Hansard, Official Report, 1 Dec 1954, Vol. 535, c. 176–7
27. AVIA 65/1117 Encl 53, National Archives
28. Aborigines and Atomic Testing in South Australia, Kingsley Palmer, No date
29. Admiralty message to CinC Mediterranean, Home, and Far East fleets, December 15 1955
30. https://damned.mirror.co.uk/img/docs/diana-captain-log.pdf
31. The Hazards to Man of Nuclear and Allied Radiations, 1956, HMSO
32. Testimony to the Royal Commission by Leonard Perkins, UPI archives, 4 Jan, 1985
33. Office of the Commodore, Special Squadron, 4 Mar 1956
34. Testimony to the Royal Commission, Feb 1 1985, p4953
35. ADM1/26714, Operation MOSAIC - stationing of *DIANA* for scientific observation of fallout, March 13 1956
36. *The Argus*, Melbourne, 15 Aug 1956, National Library of Australia
37. Letter from Penney to Sir Frederick Brundrett, MoD, Dec 22 1955
38. Buffalo Trials Indoctrination Programme, notes of meeting held at War Office, 15 Aug 1955
39. Gp Capt SWB Menaul, DDOps, 29 Nov 1955 internal memo
40. AVIA 65/1131, National Archives
41. DEFE 16/551, National Archives
42. British Nuclear Tests in Australia - Royal Commission, 20 November 1985 - Volume 1, 6.4.51
43. Op Buffalo contaminated clothing trials, summary of instructions, Nov 1956
44. AWRE report No T 11/57 The Dose Received at Various Parts of the Body by a Man Walking Over Contaminated Ground, CF Barnaby, April 1957
45. Maralinga: The Anangu Story, by the Oak Valley and Yalata communities, with Christobel Mattingley, No date, pp 44-45
46. British Nuclear Tests in Australia - Royal Commission, 20 November 1985 - Volume 2, 10.2.17

47. No Conceivable Injury, Robert Milliken, 1986
48. https://exposure.press/the-pacific-venture-operation-bagpipes/
49. AVIA 65/786, Reports on Operation Bagpipes, 1954, National Archives
50. CAB 134/940, comments on a report outlining the defence implications of a Hydrogen Bomb attack recorded by Norman Brook at the Cabinet Office, 25 Mar 1955
51. COS (55) 105th meeting, confidential annexe, 20 Dec 1955
52. Christmas Island Cracker: Account of the Planning and Execution of the British Thermonuclear Bomb Tests, 1957, by Wilfred Oulton
53. Hansard, Official Report, 7 Jun 1956, vol 553 cc1283-6
54. Christmas Island Cracker, Oulton, p333
55. Fissionline 67, 23 Mar 2023, pp 6-7
56. AIR 2/13726, Report on a Visit to Operation Grapple by Sqn Ldr SJ Pooley, National Archives
57. British Nuclear Tests in Australia, Royal Commission, 20 November 1985 - Volume 1, 9.0.5
58. Op Antler, appendices to report on air operations, National Archives
59. Memories of nuclear sampling and tracking, Australia and Christmas Island, 1956-7, Sqn Ldr JF Spatcher, presented to High Court Nov 2008
60. Atomic Weapons Tests Safety Committee, report to the Prime Minister on the Antler Trials, 21 Nov 1957
61. Hansard, Official Report, 8 Feb 1983, col 293
62. AB 16/2371, Grapple X committee papers, National Archives
63. FO 371/129281/212/675, The National Archives
64. AB 16/2371, Grapple X committee papers, National Archives
65. *Sunday Mirror*, 23 Feb 1958, pp 12 - 13
66. Hansard, Official Report, 14 Feb 1958, col 820
67. *Western Mail*, 28 Aug 1957, p7
68. *Sunday Mirror*, January 25, 2009, p22
69. GY 6/RC, Grapple Y interim report, National Archives
70. Increase in Cancer Risk for Japanese Workers Accidentally Exposed to Plutonium, by Ed Lyman, Director Nuclear Power Safety, Union of Concerned Scientists, 9 Jun 2017

Fallout

1. British Nuclear Tests in Australia - Royal Commission, 20 November 1985 - Volume 1, 8.4.74
2. Maralinga: The Anangu Story, by the Oak Valley and Yalata communities, with Christobel Mattingley, No date, p12

3. British Nuclear Tests in Australia - Royal Commission, 20 November 1985 - Volume 1, 8.4.95
4. British Nuclear Tests in Australia - Royal Commission, 20 November 1985 - Volume 1, 8.4.97
5. ARPANSA report, https://web.archive.org/web/20080831115407/http://arpansa.gov.au//pubs/basics/maralinga.pdf
6. Genetic Effects of Irradiation with Reference to Man, by D. G. Catcheside, 2 Jun 1947, prepared for the Tolerance Dose Panel of the Protection SubCommittee of the Medical Research Council
7. Hansard, Official Report, 21 May 2019, col 699
8. British Nuclear Tests in Australia - Royal Commission, 20 November 1985 - Volume 1, 5.5.33
9. https://www.armscontrol.org/factsheets/nucleartesttally
10. National Archives of Australia, Maralinga Cemetery Records, item ID 1900550
11. A.E. Oldbury, "Report on Final Radiological Survey and Decontamination Operations at Christmas Island prior to the Closing Down of the Base on 30th June 1964", (1964), Atomic Weapons Research Establishment, UK Atomic Energy Authority.
12. British Nuclear Tests in Australia - A Review of Operation Safety Measures and of Possible After-Effects, Australian Ionising Radiation Advisory Council, January 1983
13. Hansard, Official Report, 18 January 1983; Vol. 35, c. 166
14. Letter from A.A. Garran, assistant secretary, Department of Resources and Energy, to Mr G. McGregor, First Assistant Secretary, Uranium and Nuclear Division, 16 March 1984
15. The report of the Royal Commission into British nuclear tests in Australia, conclusions and recommendations, p26
16. Mortality and cancer incidence in UK participants in UK atmospheric nuclear weapon tests and experimental programmes, Darby, Kendall et al, January 1988
17. https://www.reaganlibrary.gov/archives/speech/statement-signing-radiation-exposed-veterans-compensation-act-1988
18. McGinley & Egan v United Kingdom, ECHR, 1999
19. 10 September, 1997
20. National Archives, DEFE 24/3561
21. https://archivedproceedings.econference.io/wmsym/2009/pdfs/9526.pdf
22. More than double the annual background dose in the UK
23. https://www.abc.net.au/news/2012-03-29/green-dumping-on-traditional-owners-the-ugly-face-racism/3919296
24. https://damned.mirror.co.uk/img/docs/maralinga.pdf

25. https://www.bailii.org/cgi-bin/format.cgi?doc=/ew/cases/EWHC/QB/2009/1225.html
26. Grappling with the Bomb: Britain's Pacific H-bomb Tests, Nic Maclellan, 2017
27. Speech by Ambassador Makurita Baaro, Informal Meeting of the United Nations General Assembly to mark the 2015 Observance of the International Day against Nuclear Tests, UN Headquarters, New York, 10 September 2015.
28. Australia's nuclear tourist hotspot, *Australian Geographic*, 14 Sept 2016
29. Testimony to the Royal Commission, 15 Jan 1985, p4439
30. The nature of Pu-bearing particles from the Maralinga nuclear testing site, Australia, Scientific Reports, Cook et al, 21 May 2021

Exposure

1. Dogs and Lampposts, Richard Stott, Date missing, pp. 126, 231.
2. *The People*, Sunday 28 December 1997, page 4
3. DEFE 19/598, veteran correspondence, National Archives
4. Project 4.1 Final Report, "Study of Response of Human Beings Accidentally Exposed to Significant Fallout Radiation", Cronkite et al, Naval Medical Research Institute, October 1954
5. National Archives, DEFE 68/1349
6. MAF 250/238, Home Defence Committee: papers and minutes of Strath Report, 1955
7. Leaked report by Committee Examining Radiation Risks from Internal Emitters, New Scientist, July 18, 2004
8. *Sunday Mirror*, 21 October 2007, p24
9. *Sunday Mirror*, 18 November 2007, p2
10. *Sunday Mirror*, December 2, 2007, p28
11. *Sunday Mirror*, January 13, 2008, p33
12. *Sunday Mirror*, April 20, 2008, p27
13. *Sunday Mirror*, June 22, 2008, p22
14. *Sunday Mirror*, August 3, 2008, p28
15. British Nuclear Test Veterans Health Needs Audit, Miles and Green Associates, October 2011
16. Hansard, July 4, 2014
17. A Report on the Health effects of Low-Level Radiation Sustained as a Result of the Nuclear Weapons Testing Program Conducted by the US Government, 1980
18. https://damned.mirror.co.uk/img/docs/warprension-doc.pdf
19. *Daily Mirror*, April 9, 2018, p20

20. *Daily Mirror*, April 23, 2018, pp 22,23
21. *Daily Mirror*, May 28, 2018, pp 6,7
22. Ministry of Defence FOI 2024/02835, April 11, 2024
23. https://www.theguardian.com/world/2018/dec/23/british-nuclear-archive-files-withdrawn-without-explanation
24. *Daily Mirror*, April 4, 2019, p19
25. Hansard, Official Report, 11 Jun 1958, col 23
26. https://www.labrats.international/post/uk-medal-campaign-request-denied?postId=5fd115bb99ea360017bfbd93
27. *Daily Mirror*, December 16, 2021, p22
28. https://www.mirror.co.uk/news/politics/andy-burnham-calls-boris-johnson-26282818
29. Mortality and cancer incidence 1952–2017 in United Kingdom participants in the United Kingdom's atmospheric nuclear weapon tests and experimental programmes, Michael Gillies and Richard G E Haylock, February 23, 2022
30. Hansard, UIN 190324, June 20, 2023
31. Mortality and cancer incidence in UK participants in UK atmospheric nuclear weapon tests and experimental programmes, NRPB Report R214, Darby, Kendall et al, 1988
32. https://www.mirror.co.uk/news/politics/mod-lied-years-zero-radiation-31957784
33. https://x.com/fleetstreetfox/status/1704509420753813849
34. https://www.independent.co.uk/climate-change/news/boris-johnson-raf-plane-photo-b2026043.html
35. Hansard, Written Answer UIN 200586, tabled 10 Dec 2018
36. *Daily Mirror*, April 18, 2022, p24
37. https://www.labrats.international/bombshell
38. King's College London, DEFE 26/551
39. ADM1/26714 - HMS *Diana* stationing, National Archives
40. https://www.mirror.co.uk/news/politics/boris-johnson-delivers-plutonium-jubilee-27915173
41. PM's letter to Veterans of the UK's Nuclear Testing Programme, 5 Sept 2022
42. Hansard, Official Report, 7 Nov 2022, col 5
43. Hansard, Written Answer UIN 77326, tabled on 2 Nov 2022
44. FOI to AWE, ref FOI2024-004
45. Hansard, November 21, 2023
46. https://www.mirror.co.uk/news/politics/proof-mod-lied-nuked-blood-32944207